Frederick Guttmann Ramirez

THE ABYSS

First edition. April 9, 2024.

ISBN: 979-8231491643

Written by Frederick Guttmann.

The Rebellion of Sakla

I

The Abyss

Project Magen

Frederick Guttmann R.
www.frederickguttmann.com[1]
The Abyss – The Rebellion of Sakla
(2012-2013)
172 pages
Cover: **Aday Quintero P.**

1. http://www.frederickguttmann.com

CONTENT

INTRODUCTION

"When we look back through the historical time of patriarchy... there seems to be some terrible inevitability, a relentless desire to shatter feminine essence, human and divine ..." Leonard Schlain (The Goddess Alphabet)

Thanks to previous works we have prepared the way for this investigative work and presentation on the origins of the world and the forces that act in it. In the book "Creation vs. Evolution 2012", we were already preparing ground by dealing with issues of the past in relation to archaeological evidence that shows that man did not come from apes. Also in line with said matter I compiled substantial information about UFOs in "Wandering Stars", since the beginning of the creation does indeed come from outside. Likewise, those who have read our 2011 work, "Armageddon, E-5", will have been intrigued as to how was the origin of all this universal conflict mentioned there? Well, we are entering hot topic that many have tried to gather, blindfolded on complex, delicate, uncertain, dark and incomplete issues.

Everything that exists can be very different from what we believe it to be and from what we have been popularly taught. We would need to experience an Eternal Life to know all the mysteries of this unreal universe in which we live and to understand the cosmos. Many, seeing the limitations of physical laws to explain the universe, have been forced to create new sciences such as "quantum physics" or "quantum mechanics." For example, in "Wandering Stars" he clarified doubts about the UFO issue, since it is what catapults us to talk about these issues, despite the fact that the definition of "extraterrestrial" will now fall infinitely short.

Man wonders, at least once in his life, about his origin. Logically, if we don't know where we come from, we can hardly determine where we are going, who we are, or why we are here and now. We see the appearance of human beings in the fossil record, which would be hundreds of millions of years old, an aspect that suggests that there were humans on the surface of this planet in prehistory, and that due to events that we will now disclose, it was lost. recording of these events. So, the evolutionary theory is only a topic and a way of seeing life, not a scientific truth, and what is a scientific fact is to clarify the idea that man has practically always been man. Although if so, when did it appear? And where? Are we originally from this orb?

As we will remember in "Creation vs. Evolution", prehistoric tools and artifacts have been found that seem to be out of time. This shows us that there were other civilizations before the age of the cavemen, which had more technology than their apparent predecessors. Seen that way, this significantly changes human history in a radical way. Thinking that they could be beings from other planets is partly plausible in some cases, but not in the ones mentioned above, since some creatures that traveled through the sidereal universe at speeds greater than light would not use arrows, nails, screws, etc. hammers or spears, unless they got lost here. Even with everything, who says that they came from other planets and not from other dimensions, realities or universes? We must also accommodate the fact that suggests the possibility that humanity had "fractures" in its social development, which gave birth at different ages, losing the historical and intellectual record they had previously, over and over again.

Who were these people and how does this evidence lead us to the elementary answer? Where did the man come from? Why is there death, suffering and evils? The evidence tells us, at least about us, that we simply "reemerged" from the caves some 6,000 years ago, and that everything we know of already existed in the past. Perhaps by meddling in the synthesis that archaeological findings expose us, we may find the long-awaited response from our annals, or by unearthing ancient texts: «*What was it? The same as it will be. what do what has been done? The same thing that will be done; and there is nothing new under the sun. Is there anything that can be said: Behold this is new? It was in the centuries that have preceded us. There is no memory of what preceded, nor of what will happen there will be memory in those who will be after.*» (King Solomon. Book of Ecclesiastes 1:9-11. TANAQ)

The truth is out there, you just have to dust it off. The basis of our focus today is first to investigate and then show what was "the beginning", what is the origin of man based on history, ancient texts and archaeology, not that chimerical concept that we have been led to believe. for centuries the Romans who absorbed Christianity, or for 250 British Freemasonry with naturalistic ideas. We also want to undermine the controversial and mysterious Lucifer rebellion and its consequences. The oldest references that mention the appearance of the universe are mentioned by various cultures on Earth, and therefore, these, along with the evidence unearthed from the bowels of this globe, will be our main weapons in the face of ignorance of this mysterious bozo. which is the primal beginning, the demons, the giants, the gods, man and creation itself. Some of the most striking accounts represent the basis of Gnosticism, and these are the Nag Hammadi scrolls (Egypt), but we also have data from Qumran (Dead Sea, Israel), unearthed by both groups from 1945 and 1947. However, we are not going to enter with "solid food" only, but we will exhume the fossil findings that will put together the puzzle of our history.

With the staging of this book we want to clarify, among other key points, the explanation of what tends to be called "mythology" and "legend", as well as the possible meaning of so many artifacts out of time and, of course, put the magnifying glass on the famous book of Genesis and its creationist version. I will repeat once again that this work takes us on a journey through time, unraveling information hidden from the public, and highlights another that has not been "seen" even by the most learned. If we look at known history or what we have been led to believe, there are innumerable links in our humanity: When did we appear? What was it really like in the beginning? Who were Adam and Eve for sure? How did it all start? These questions and many more it may be time for them to be resolved, and the "veil removed", at least to the extent possible, since in this age "we know in part, and in part we prophesy; *but when the perfect comes, then what is in part will end.* » (1st Letter of Paul to the Corinthians, 13:9-10. New Testament, the Bible).

Our past is the spitting image of our future. As we read shortly before in the book of Ecclesiastes, there King Solomon exposes us that everything that man believes he has discovered is nothing more than something that "already existed in the past." Today, thanks to technology and great people who fought to make discoveries known that our minds today can objectively visualize, we have achieved a certain understanding or "understanding" superior to all the generations that have preceded us. Our scientific growth allows us to look at the sky and understand countless things that our ancestors could not have even imagined in their entire lives. The advances of the last 70 years have never been seen in the entire "known" history of the human race: The trip to the Moon in 1969, the beginning of the nuclear age in 1945, telecommunications, chips and all the world events that we let slip by unnoticed; things that were previously impossible to imagine:

a submarine trip to the bottom of the oceans, at the time it was unthinkable; conquering the airspace was something only comparable to the fantasies and dreams of the craziest men. Now science fiction no longer exists.

This generation has known, is knowing, and is going to know certain and precise things that all the prophets recounted at the time: " *Truly I tell you, this generation will not pass away until all this happens.*" (Matthew 24:34, Mark 13:30 and Luke 21:32). Thus, all the events that have occurred are of a transcendental nature for the great outcome that has been spoken of in many cultures since millenary eras, and that we have qualified in "Armageddon E-5" and "Recognizing the Time of the End". All these prophecies are concerning the days that we are living. At the same time, what we see and have registered in recent history is nothing more than the repetition of events that occurred in the past, in ages whose records have practically disappeared. So who is right, science or religion? Maybe neither of the two, and at the same time both.

In "Creation vs. Evolution" he questioned the supposed most important branch that exposes the descent of humanity and part of the "world" (as the "Living Universe" is designated in the Hebrew and Greek terms: the "genesis" and "evolution", that is, the approaches with the greatest weight at a global level), and now we will do the same with the Creationist Theory of mythologies, cosmogony. Many claim to be right, all religions say that theirs is the true one, the human justifies himself and is convinced and debased in his reasoning, but let the accumulation of documented points speak for themselves. Humanity itself will explain its origin, and in this regard we will also take a look at the information compiled by various countries in relation to what they consider to be its origins -although sadly for some it is not history but remains in the drawer of mythologies, also due to the passing of the years and the sum of stories and fanciful additions to fatten the folklore.

Let us bear in mind that Mythology is what we could call "history" fused or intermingled with "legend", sometimes very fanciful, other times more real, but based on a fact that obviously happened, but in all cases it must be previously Compared with reality, with history and with science, because over the years they come to lose credibility due to the loss of their sources and urban tales. It is considered a myth because there are no official records and because the stories have become irreconcilable novels or legends with modern logic. However, all data can help us in this important clarification. I invite you to immerse yourself in this journey that can change your mind, as it has changed mine.

I.
THE ABYSS

THE HOLLOW EARTH

"The Earth and I are one mind."
(Chief Josephus of the Nez Perce)

The ancient world had firm beliefs that beneath the Earth was a giant hole of colossal proportions. In that dark and abysmal place there would be gloomy and evil beings from time immemorial, some swarming and others imprisoned in dungeons. Many legends hang over this place and its inhabitants, including fables about its origin and relationship with the beginnings of the world. For its part, modern science teaches that the center of our planet is an enormous solid mass of nickel, iron, and other minor metal compounds in both the Inner Core and the Outer Core. The peoples of old, one could say all of them, saw something very different: they took it for granted that another world was developing below the surface. The book of Exodus says: " *You shall not make for yourself an image, nor any image of anything that is in the heavens above, nor on the earth below, nor in the waters under the earth."* (Vers. 5:8) What could be under the waters that could be a problem for God, worshiping the Israelites? The theory of Planetary Differentiation could be incorrect if an "artificial" process could be attributed to the formation of the planet on which we live and to the rest of the spheres of space. Seismic measurements show schemes that today can be refuted since they themselves reflect a "void" at a depth of 400km, as a kind of "change of gravitational attraction", and then, another 400km below, THERE IS NOTHING. Studies in the US on this issue have not received further funding and no one wants to talk about it. It is forbidden to fly over the poles, the Google Earth system censors photos of the polar zones -both of the Earth and of the planets and moons of the solar system. People who claim to have proof that the Earth is indeed "hollow" are also silenced. This is not a conspiracy theory or a laughable tale, but a fact that contradicts most accepted hypotheses about the origin and formation of the Earth. The theory that we have been taught about the formation of the stars –cold or hot- is completely false. Its composition is more like that of a "hollow ball", with two important openings at the poles. These concepts can explain the meaning of so many ancient myths about the creation of the planet, where a void would be added inside, which was biblically referred to as the Abyss.

The Egyptians considered the existence of the world below, the Duat, for which they created long underground galleries, and they ran into other existing ones (they also believed in the Halls of Amenti below the surface). The ancient peoples of America and Tibet knew a lot about the kingdoms below the Earth and about their great teachers. The Greeks regularly spoke of the entrances to Hades and how brave heroes tried to get there. This place was defined by the Hebrews as Sheol (sheol). In the Mesopotamian culture there are records of trips to the underworld, especially the Gilgamesh novel. For the Norse the world below, or Hell, was a step away from existence. From there they took the heroes that the Valkyries would gather in Walhalla for the Rangarök. To the high priest Melchizedek they would have said: " *But those in heaven spoke many words, along with those on earth, and under the earth.*" (Melchizedek. Nag Hammadi Library) Other parts of the Hebrew and apostolic scriptures say: «... *so that in the name of Jesus every knee should bow, of those who are in heaven, and on earth, and under the earth* ...» (Philippians 2:10) Paul himself seemed not to be unaware of the existence of peoples below the world, without confusing the area of the deceased, also established below. John wrote: " *And no one, neither in heaven nor on earth nor under the earth, could open the book, nor even look at it.*" (Revelation 5:3) And further on he added: "*And to every creature that is in heaven, and on the earth, and under the earth, and in the sea, and to all things that are in them...*" (Revelation 5:13) An example, in this case about the vastness of the world below -although there is much more material about this- and its pillars, we see: « *Thus says the Lord: If the heavens above can be measured, and explored below the foundations of the earth ...*» (Jeremiah 31:37) Enoch spoke substantially about it, but looking at the book of Job it seems that we find a reference to the polar openings: « *From what womb did the ice come out? And the frost in the sky, who begat it? The waters harden like stone, and the face of the deep freezes.*" (Job 38:29-30)

However, more recently, according to eminences of the 17th century, in the center of the Earth, below it, shines a central sun that remains static, called by ancient cultures as "the black sun" or the "emerald sun" (for the Egyptians the deity Atum was the personification of said internal star). Edmund Halley –writer of more than 50 books- (1656 - 1742), discoverer of the comet that bears his name (Halley's Comet) and royal astronomer of England in the 18th century, drew up a scheme where the Earth appeared as hollow in its nucleus and below there was life as we know it on our surface. He also believed that this planet, being hollow, housed 3 plants inside. His theory was based on the fact that the Earth was made up of several layers or "levels" – like an onion. The respected Swiss mathematician and physicist Leonard Euler (1707-1783) said that the Earth was a shell that had two openings at the poles, a central sun, and was inhabited. In this order, Cyrus Teed also had a theory that expressed that there would be life under the surface of the Earth and they would have a central sun down there, natural environmental conditions and natural gravity, that is, similar to those on our surface. We see that the scientific world was not immune to this theory, and they alternated with various works of fiction on the same subject, the most important of which were: The Adventures of Arthur Gordon Pym, by Edgar Allan Poe (1833), in which the hero and his companion have a terrifying encounter with beings from the interior of the Earth; and Jules Verne's Journey to the Center of the Earth (1864), in which an adventurous professor, his nephew, and a guide penetrate the interior of the Earth through an extinct volcano in Iceland, and find new skies, seas, and gigantic and prehistoric reptiles that swarm in the forests. The belief in a hollow Earth was so widespread that even Edgar Rice Burroughs, the famous author of Tarzan, felt compelled to write "Tarzan in the Bowels of the Earth" (1929), in which the famous son of the jungle goes to Pellucidar, a world that lies on the inner surface of the Earth and is lit by a central sun. HP Lovecraft's "The Shadow Beyond Time" (1936) transported the theme to the present day by describing an ancient, subterranean race that dominated the Earth 150 million years ago and, ever since, in the refuge of Inner Earth, he has invented airplanes and atomic vehicles, and masters time travel and extrasensory perception.

New theories about UFOs related them to this intraterrestrial world. These theories led to the recovery of the legends of the "lost" civilizations of Atlantis and Thule, in the belief that the latter was in the Arctic (not to be confused with Dundas, formerly Thule, the Eskimo enclave in Greenland, which is today a US air base and communications center). However, it was also believed that another possible source of origin of UFOs could be related to Antarctica. This theory arose from the publication of John G. Fuller's convincing book, "The Interrupted Journey" (1966), in which the author tells the story of Betty and Barney Hill, an American couple who, during psychiatric treatment due to an inexplicable period of amnesia, he remembered under hypnosis that they had been abducted by aliens, examined inside a flying saucer and informed that the aliens had bases all over the Earth; some at the bottom of the sea and at least one in Antarctica. The British Geological Society said at the time: "*There are dozens of elements against the possibility of the Earth being hollow, yet there were hundreds of elements against the possibility of a Solid Earth.*" Let us also keep in mind that « *volcanic activity does not begin at 6,500km, as they say. If a force of this nature existed within the Earth, nothing would be left alive on its surface. According to some volcanologists, magmatic activity does not begin beyond 30km depth. If we look at the possible mass, if this Earth were solid, we would be 1 cm high due to gravitational pressure, this would be enormous. We only have to take a calculator and calculate the mass to which we would be subjected and the gravity by effect to which we would have to be subjected.*» (UnknownWorld.com). Sir John Leslie (1766-1832), physicist and inventor, native of Edinburgh, inventor of the differential thermometer in the early 19th century, said that the Earth was hollow. In Classical Greece, at Socrates' trial, Meleo accuses him of constantly talking about the World of the Below, because he found such statements uncomfortable.

Athanasius Kircher (1602 - 1680) talks about a book published in the 17th century that talks about the Underworld. A. Kircher was born in Geisa, Germany, and was a Jesuit priest, polyglot, scholar, orientalist scholar, with an encyclopedic spirit and one of the most important scientists of the Baroque; In his book "Visa for Another Earth", Jacques Bergier deals extensively with this subject of the intraterrestrial world. Some scientists whose name has not been revealed affirmed a few years ago, in defense of these theories, that the magnetic pole is not a point, but a perfect circular line, the same in the North as in the South, and that any point of this circular line could be identified with the Pole. If this is true, it would explain what happened in 1909, when Dr. Frederick Cook was acclaimed as the conqueror of the North Pole, a fact that took place on April 21 of that year. Days later, Vice Admiral Robert E. Peary would declare that he was at the North Pole on the 6th and that he planted an American flag there. Neither of the two explorers had witnesses to their feat. Both were sure that they had reached the North Pole. Cook must have seen the flag left by Peary, having been planted by him. But it didn't show up anywhere. *"There lives the superior race, the same one that will one day rise to annihilate us"*, defended the Englishman Bulwer Lytton two centuries ago, and his theory would be accepted by the philosophers of Nazism, who would be in favor of the existence of an inner sun. US NAVY Admiral Richard E. Byrd stated in 1947 that *"the United States could be attacked by enemies that would come from beyond the poles."* Walter Sullivan (writer of "In Search of a Continent" - 1928), who was a chronicler of the expeditions of this renowned polar explorer (RE Byrd), said that we could be attacked by "more advanced ships that would come from the polar regions. "The same was said in 1947 by the General of the USAF (United States Air Forces), George Churchill Kenney (1889 - 1977), who participated in World War II; by Rear Admiral Richard Cruzen, head of the polar expeditions led by Admiral Byrd; and by Rodolfo Panzani, Rear Admiral of the Argentine Navy, who referred to these words in his book "La Naturaleza de la Antártida".

While still young, Olaf Jansen traveled north with his father and arrived in a country where the temperature was pleasant and the sun misty, different from what travelers knew. They toured the place, populated by exuberant vegetation, and went to find themselves in front of a real mammoth. On the way home, the boat hit an iceberg and Mr. Jansen fell overboard. His son would be rescued by the crew of another ship, who heard his amazing story. Consequently, Olaf was locked up in an asylum, where he remained for 20 years. He had time to write a book that he called

"The smoking god", referring to the strange sun that seemed to give off smoke. About Richard Byrd it is said that when he returned home he was severely reprimanded, and forbade him to make any more statements typical of a madman. Byrd would die months later, disappointed to see his compatriots refuse to accept what he considered the most sensational discovery in recent years. However, not everyone was going to call him crazy. The Italian Amadeo Giannini meditated on Byrd's words and carried out research on some testimonies from the last century and ended up writing a curious work entitled Worlds beyond the Pole, published in 1959. Giannini emphasized in his book the adventure of the Norwegian Fridtof Nansen, who discovered on August 3, 1894 in the northern region of Greenland something that perplexed him: he was at about 86° north latitude when he found some logs dragged by the sea current coming from the north. This finding would be confirmed shortly after by Commander McClure when he was exploring the land of Banks, an island in the Arctic archipelago. And also by Captain Beechy on the west coast of Spitzberg Island, who found wild geese inexplicably flying north at the same time. Another similar piece of information was that of a mammoth found in Siberia in 1799.

In those days, the American William Reed would launch a curious theory about the mysterious polar region, based on the book written by the crazy Norwegian. He said that in each of the two Poles of the Earth a circular opening opens that allows entry into the interior of the Earth. Had this man also gone mad? Reed explained that the force of gravity is so intense at the polar openings that the water from the interior rivers rushes to the surface of the planet, where it freezes and becomes icebergs. These are then broken into gigantic fragments, producing the strange tidal waves well known to polar explorers. Inside the blocks of ice, plants and animals of all sizes reach the surface, such as a mammoth found in Siberia in 1799. Scientists at the time declared that the mammoth had been encased in ice for thousands of years. Reed claimed, on the contrary, that the huge animal unwisely ventured into the opening of the outside world and was carried by the current to the Siberian lands. In 1902 another mammoth would appear, in the Bereskova region, and recently the primitive being that floated wrapped in ice off the Aleutian Islands. The scientists affirmed that, if both had been frozen under normal circumstances, at not too low temperatures, the decomposition process should have continued. That is to say, that the age of the mammoth and of the human being was not superior to one or two centuries. Does this mean that Reed's theory might be true after all and that early mammoths and humans came from inside the Earth? William Reed would have a strong defender of his ideas, from 1920, in the character of his compatriot Marshall B. Gardner, who had observed certain strange phenomena and devised new theories to complete the previous ones. He had observed the colored snow that occasionally fell over arctic regions, and wondered if it might not be caused by pollen from plants growing in that unknown interior of the planet. Gardner looked at the polar cap on Mars and wondered, too, whether it might not be just as hollow. He said that the whitish color that characterizes the Martian pole is not due to ice, but to the clouds that pile up near the opening. If the polar cap sometimes vanishes, Gardner added, it is because the clouds are introduced through the opening, when summer arrives, as if it were a gigantic tide whose mechanism has not yet been understood.

Although it sounds absurd, these theories are as old as humanity and have even received the approval of an almost contemporary government. In some ancient legends there is talk of the underground kingdom of Agarthi, which is located under the mountains of Tibet. It has been said on many occasions that the much-discussed UFOs do not come from space, but rather have their bases inside the Earth, which they leave by exiting precisely through the two openings that exist at both Poles. "The superior race lives there, the same one that will one day rise to annihilate us", defended the Englishman Bulwer Lytton two centuries ago, and his theory would be accepted by the philosophers of Nazism, who would be in favor of the existence of an inner sun. According to his hypotheses, this sun would illuminate a hollow earth whose inhabitants are of the Aryan race and hate to death those of us who live on the surface of the planet, enjoying the delights of an external sun that should not hide like the other. That green earth would have been transported from its place and the axis of the planet Earth would have been modified. According to science, throughout terrestrial history some 14 massive destructions have been recorded, and where various factors operated, that is, they were precipitated by biological causes, changes in the planet's dynamics, asteroid impacts, but the most disastrous have been the oldest. For this reason the geography of our globe changed so much, since the magnetic pole was not where it is today. At present we can see that the hypothesis that there is no land at the North Pole is totally false, since in the 90s a dinosaur was discovered in the Arctic. Something similar also happened at the South Pole, where a new dinosaur was discovered. Thus we know that Antarctica already 190 million years ago was green land (hence the name Green-Land or Greenland). Millions of years later, a group of paleontologists discovered the dinosaur under the ice, near the Beardmore Glacier, at an altitude of about 4,000m. Only now this dinosaur could be identified and named, after a decade of patient studies. Paleontologists, led by William Hammer, discovered a foot, paw, and ankle on Mount Kirkpatrick.

They believe that this Antarctic dinosaur would have been about 6m long and weighed between 4 and 6 tons. Those in charge of the Glacialisaurus Hammeri study, Diego Pol, a paleontologist at the Museo Egidio Feruglio in Argentina, and colleagues determined that this new genus and species of dinosaur was from the Jurassic period. It is a sauropod, from the group of the largest dinosaurs that walked the earth, which had not previously been discovered in Antarctica. Sauropodomorphs had long necks and were herbivorous. They were distributed throughout the world, since they are found today from China, South Africa, South America, North America, and even in Antarctica. And this is because in those times the continents were still united in the supercontinent Gondwana, and the climate was more or less the same in almost all the areas that are now so separated from each other. (Sources: LiveScience, EurekAalert. Posted by Martín Cagliani. December 12, 2007) Palm trees proliferated in the Arctic 53.5 million years ago, during a temperate transition period known as the Eocene 2 Thermal Maximum, according to a study by The University of Utrecht in the Netherlands that is published in the online edition of the journal Nature Geoscience. The researchers explain that the presence of these plants indicates that winter temperatures on the continents of the Arctic region were, on average, above 8° Celsius. The scientists, led by Appy Sluijs, used marine sediments collected from the Arctic Ocean to assess environmental changes associated with the rapid warming that occurred during the Eocene 2 Thermal Maximum. This climate event is often attributed to a rapid rise in atmospheric carbon concentrations. Their model reconstruction of sea surface temperatures reached 27° Celsius, between 3° and 5° above ambient conditions. The scientists explain that the presence of palm pollen in marine sediments revealed that palm plants were present in the northernmost latitudes of the planet.

Byrd's Journal

In 1956 Richard E. Byrd returned from traveling to the poles and stated that he had seen " *a marvelous continent* " -since at the

North Pole, specifically, it is stated that there is only ice, not land-, in his own diary he states: "*I must write this diary secretly and in absolute secrecy. It refers to my Arctic flight on February 19, 1947. There will come a time in which the rationality of men must dissolve into nothingness and then the inevitability of The Truth must be accepted. I do not have the freedom to divulge the documentation that follows, perhaps it will never see the light of day, but I must, in any case, do my duty and relate it here with the hope that one day everyone will be able to read it, in a world where selfishness and the greed of certain men will no longer be able to suppress The Truth.*" His words are not strange, since a not very old film has been revealed, made from an airplane, where mountains are seen in the Arctic. However, in Byrd's diary he continues his entry, based on that day's logbook: ...«*Expanses of ice and snow below us, seen yellowish coloration with line drawings. Altered the route for a better examination of these colored configurations, also seen purple and pink colorations.*" "*Both the magnetic compass and the gyroscopic needle begin to rotate and oscillate, it is not possible for us to maintain our route with the instruments. We point the direction with the solar compass, everything still seems in order. Controls seem slow to respond and operate, but no indication of icing.*" ..."*29 minutes of flight since first sighting of mountains, not hallucination. It is a small range of mountains that we have never seen before.*" Then he says: « *Besides the mountains there is something that seems to be a valley with a small river or stream that runs towards the central part. There should be no green valley down here! There is something decidedly strange and abnormal here! We should fly over only ice and snow! To the left are large forests on the slopes of the mountains. Our navigation instruments are still spinning like crazy.*" ..."*I alter altitude to 1400 feet and make a full left turn to better examine the valley below. It is green with moss and very bushy grass. The light looks different here. I am not able to see the Sun. We take another left turn and sight what appears to be some kind of large animal. It looks like an elephant! NO!!!. It seems to be a mammoth! Is incredible! However it is so! We descend to 1000 feet and I use a binocular to better examine the animal. It is confirmed, it is an animal similar to the mammoth."*

«We found other green hills. The outside temperature indicator shows 24° Celsius. Now we continue on our route. On-board instruments now appear normal. I am puzzled by their reactions. I'm trying to contact base camp. Radio doesn't work." ..."The surrounding landscape appears level and normal. In front of us we can see what appears to be a city!!! It is impossible!. The plane seems light and strangely buoyant. The controls refuse to respond! My God!. To our right and to our left are devices of a strange type. They approach and something radiates from them. Now they are close enough to see their badges. It is a strange symbol. Where we are? What happened?. Once again I pull firmly on the controls. They don't respond!!! We are firmly trapped by a kind of invisible steel trap.» Then he prays: " *Our radio squawks and there comes a voice speaking English with an accent that sounds decidedly Nordic or German! The message is: - Welcome to our territory, Admiral. We'll land you exactly in seven minutes. Relax, Admiral, you are in good hands -. I notice that our plane's engines are out. The device is under strange control and now it turns on its own". ..."We received another radio message. We are beginning the landing maneuver and shortly the plane vibrates slightly beginning to descend as if supported by a huge, invisible elevator. Some men are approaching, on foot, the plane. They are tall and have blonde hair. In the distance is a great glittering city, vibrant with the colors of the rainbow. I don't know what will happen now, but I don't see any traces of weapons on the approaching ones. Now I hear a voice ordering me, calling me by name, to open the door. I execute.»* After these notes, taken from the "board log", the Admiral writes down what happens: *« From this point on I write down the events that follow, calling them back to memory. This settles the imagination and would*

seem crazy if it had not really happened. The technician and I were taken off the plane and warmly welcomed. Then we were embarked on a small means of transport similar to a platform but without wheels. He led us towards the glittering city with extreme speed. As we approached, the city seemed made of glass. In a short time we reached a large building, of a style I had never seen before. It seemed to have come from Frank Lloyd Wright's designs, or perhaps more precisely from a Buck Rogers scene! »

« They offered us a kind of hot drink of something I had never tasted. It was delicious. After about 10 minutes, two of our amazing hosts came to our accommodation, inviting me to follow them. He had no choice but to obey. I left my radio technician and we walked a bit until we entered what appeared to be an elevator. We descended for a few moments, the elevator stopped, and the door slid up silently. We then continued down a long corridor illuminated by a pink light that seemed to emanate from the very walls. » ... *« One of the beings signaled us to stop before a large door. Above it was an inscription that I was not able to read. The great door slid soundlessly open and I was invited inside. One of the hosts said: - Don't be afraid, Admiral, you are going to have a discussion with the Master. - I entered and my eyes slowly adjusted to the wonderful coloration that seemed to completely fill the room. Then I began to see what was around me. What was shown to my eyes was the most astonishing sight of my entire life. Indeed, it was too magnificent to be described. It was delicious. I don't think there are human terms capable of describing it, in every detail, fairly. My thoughts were gently interrupted by a warm and melodious voice: "I welcome you to our territory, Admiral."* » ... *«I saw a man with delicate features and with the signs of age on his face. He was sitting at a large table. She invited me to sit in one of the chairs. After I sat down, he linked his fingertips together and smiled. He spoke sweetly again and said the following: - We have let you in here because you are of noble character and well known*

on the surface world, Admiral. Surface world! I almost ran out of breath! "Yes - the Master emphasized with a smile - You are in the territory of the Arians, the submerged world of the Earth. We will not delay your mission much and you will be accompanied back on the surface and also without danger. But now, Admiral, I will tell you the reason for your summoning here. Our interest began exactly immediately after the explosion of the first atomic bomb by your race on Hiroshima and Nagasaki, in Japan. It was at that unsettling moment that we dispatched our means to your surface world. flyers." "The Flugelrads, to inquire into what your race had done. This is obviously past history, Admiral, but let me continue. See, we have never, before now, interfered in the wars and barbarism of your race, but now we must do it as soon as you have learned to manipulate a type of energy, the atomic, which is not, in fact, for man. Our emissaries have already delivered messages to the powers of your world and yet they do not attend to them.»

«Now you have been chosen to witness that our world exists. See, our culture and our science are thousands of years ahead of yours, Admiral." I interrupted him: "But all this that has to do with me, Sir!" The Master's eyes seemed to penetrate deep into my mind. and after studying me for a moment, he replied: "Your race has reached the point of no return, for there are some among you who would destroy your entire world rather than relinquish power as you know it..." ... "I nodded, *and the Master continued: "From 1945 onwards, we have tried to make contact with your race but our efforts have been met with hostility: our Flugelrads were fired on. Yes, they were even followed with malice and animosity by your combat planes. So now, my son, I tell you that there is a great storm on the horizon, for your world, a black fury that will not be extinguished for several years. There will be no defense with your weapons, there will be no security in your science. It will devastate until every flower of your culture has been trampled and all things human are scattered in chaos. The recent war*

has only been a prelude to what is yet to come to your race. We, here, can see it more clearly every hour... Do you think I'm wrong?" "No - I answered - it has already happened once in the past; The dark years arrived and they lasted 500 years." "Yes, my son - replied the Master - the dark years that will come now for your race will cover the Earth with a mortuary cloth, but I believe that some of you will survive the storm, more than this I don't know! We see in the distant future emerge again, from the ruins of your race, a new world in search of its legendary lost treasures and these will be here, my son, safe in our power. When the time will come we will appear to help live your culture and your race. Perhaps, by then, you will have learned the futility of war and its struggle... and after that moment a part of your culture and science will be restored to you so that your race can start anew. You, my son, must return to the Surface World with this message...".»

«With these decisive words, our meeting seemed to come to an end. For a moment I seemed to be living a dream... and yet I knew that this was reality and for some strange reason I leaned slightly, I don't know if out of respect or humility. Suddenly I realized that the two fantastic hosts, who had led me here, were again at my side. "This way, Admiral," one of Them indicated to me. I turned once more before leaving and looked at Master. A sweet smile was imprinted on her old and delicate face. "Goodbye, my son," he said to me, and made a gentle gesture with his graceful hand, a gesture of peace, and our meeting finally came to an end. We quickly left the Master's room through the large door and entered the elevator again. The door descended silently and we immediately moved up. One of my hosts spoke again: "We must now hasten, Admiral, as the Master does not wish to further delay your scheduled schedule and you must return to your race with his message." I didn't say anything, all of this was almost inconceivable, and once again my thoughts were interrupted as soon as we stopped. I entered the room and was again with my radio technician. She had *an anxious expression on her face. Walking over I said, "It's all good, Howie, it's all good." ... « The two beings pointed out to us the waiting medium, we left and soon we reached our plane. The engines were at minimum and we embarked immediately. The atmosphere, now, was charged with a certain air of urgency. When the door was closed, the plane was immediately carried up by that invisible force until we reached 2,700 feet. Two of the air assets were on our flanks, at a certain distance, causing us to glide along the return route. I must remark that the speed indicator was not indicating anything, however we were moving very quickly. ...We received a radio message: "Now we leave you, Admiral, your controls are free. Wiedersehen!!! We looked at the Flugelrads for a moment, until they disappeared in the pale blue sky. The plane seemed, suddenly, captured, by an updraft. We immediately took control. We didn't speak for a while, each one of us was immersed in his own thoughts. ...We flew over expanses of sky and snow again, about 27 minutes from base camp. We sent a Radio message, they answer us. We have normal conditions... normal. From the base camp they express relief for having established contact again." "...We landed softly at base camp. I have a mission to accomplish."*

«March 11, 1947. I just had a staff meeting at the Pentagon. I have fully recounted my discovery and the Master's message. Everything has been duly recorded. The President has been brought up to date. They hold me back for a few hours (exactly 6 hours and 39 minutes). I am carefully interrogated by the Top Security Forces and by a medical team. It's a torment!!! They put me under the tight control of the Homeland Security media of the United States of America. They remind me that I am a soldier and therefore I must obey orders.» Lastly, he notes: *« Last entry: December 30, 1956. These last few years, from 1947 to today, have not been good... Here, then, is my last entry in this singular diary. In conclusion, I must state that I have duly kept this argument secret, as ordered, for all these years. I have done this against my every principle of moral*

integrity. Now I feel the great night approaching and this secret will not die with me, but, like all truth, it will triumph. This is the only hope for mankind. I have seen the truth and it has reinvigorated my spirit giving me freedom! I have done my duty regarding the monstrous military industrial complex. Now the long night begins to approach, but there will be an epilogue. As the long night of the Antarctic ends, so the bright sun of truth will rise again and those who belong to darkness will perish in its light... For I have seen "That Land beyond the Pole, that Center of the Great Unknown". Concluded pilot Richard Evelyn Byrd. It should be noted that although this story seems fanciful, German maps and documents from World War II have been discovered where the Nazis show they have mapped land that is not on the surface, ways to access the polar openings and the like.

That hole Today

In early 1970, the Environmental Science Service Administration (ESSA), belonging to the United States Department of Commerce, provided the press with photographs of the North Pole taken by the ESSA-7 satellite on November 23, 1968. One of the photographs showed the North Pole covered by the usual cloud cover; the other, which showed the same area without clouds, revealed an immense hole where the Pole should have been, see the header photo of this article. ESSA was far from suspecting that its routine atmospheric reconnaissance photos were going to help spark one of the most sensational and celebrated controversies in UFO history. (See image at the end of this chapter). In the June 1970 issue of Flying Saucers magazine, editor and ufologist Ray Palmer reproduced the ESSA-7 satellite photos along with an article stating that the hole in the photo was real. However, it would not be the first nor the only photograph of the polar openings that would come to light. True or not, the fact is that there are testimonies, witnesses, documents, photographs and studies in this regard that weigh more than the common hypothesis taught in world education. That same man who spoke with Byrd remembers the wise teachers of whom they speak, and long ago spoke, the American Indians and the Tibetans, implying that there have been various civilizations down there for a long time.

Colonel Billie Faye Woodard wrote: "*After my arrival at Area 51, I was taken deep into the earth and did not see daylight for 11.5 years.*" This information was collected and transcribed with her permission, from a recording of a telephone communication dated January 10, 2002: "*My name is Colonel Billie Faye Woodard of the United States Air Force.*" He adds: «*I was first assigned to Area 51, in Nevada, on January 28, 1971 until 1982. In that period of service I visited the hollow interior of the earth six times, at a depth of 800 miles (approx. 1280km). Upon my arrival at Area 51 I was indoctrinated into the existence of tunnels under Area 51, and very soon after that I met several of the beings that operate the underground shuttles, who must be 13 to 14 feet tall (approx. 3 5m to 4m high). These tunnels that cross the world, have been built by a species of beings that have existed here before us, long ago.*" Then he says: « *Immediately after my arrival, I became aware of the tunnels and all the work that is carried out in the facilities themselves. I was informed that the first 15 levels of the facility were man-made, that levels 16 through 27 were already there. No one from our government did them. We were just facilitating them. My father had been assigned to Roswell. As part of my induction into the military he requested that I be assigned to him in the Pentagon. There they said "We have a new assignment for you at Area 51, Nevada, commonly designated S-4." When I entered the Pentagon I had the rank of Second Lieutenant. When I was assigned to a field commission at the Pentagon I was promoted to First Lieutenant. After 3 weeks of being there they gave me the rank of Colonel saying: "You have to be a full Colonel to be assigned to your next assignment."* » At another point he comments:

" *There were 150,000 people working in these facilities. Approximately 85% were military personnel and 15% civilian personnel. After my arrival, I was taken underground and did not see daylight for 11, 5 years.*"

Regarding tunnels below the earth's surface, Woodard said: " *The walls of the tunnels are very smooth. The walls are finished in marble, made of an impenetrable metallic substance; the surface of the walls cannot be penetrated by a diamond drill and even a laser cannot penetrate their surface. I remember there was a time when we used to see troop movements from point A to point B on the earth's surface continuously. It wasn't that long ago. Now, we hardly see it. They use these tunnels to mobilize troops over long distances. The tunnels are wide enough to drive two 18ft vehicles side by side (approx. 5m). Coming from Area 51, a shuttle leaves for the Pacific Ocean - 350 miles (approx. 560km) west of Monterrey - where there is a pyramid; another shuttle goes to the Cheyenne Mountain facility. The length of a large shuttle machine is approximately 1/4 mile (approx. 400m long). The inhabitants of the interior make use of these machines - a huge ship to move a large number of people, beings or whatever quickly.* » He adds: « *The smallest ferry is 50-60 feet in length (approx. 14-17m.), this was the class in which I traveled. The speed of the shuttles is faster than the speed of sound, they can travel from Area 51 to major sites in the interior of the earth in less than 10 Earth minutes. In 5-6 minutes you are there.*" "*The material used to build the shuttles is the same substance that the skin of the Roswell spacecraft is made of. The shuttles run on electromagnetic power using power lines from the earth. The operators I mentioned above, who are 13-14 feet (approx. 4m) tall, resemble us in appearance but are much more highly evolved, and speak via telepathy.*" "*The men either wear beards or not, and the women's skin has no wrinkles or blemishes, in fact they have a perfect fair complexion. There are seven civilizations that reside in the inner earth - which are governed by the principles of harmony. They understand and speak all the languages of the earth. His understanding of medical knowledge is phenomenal."*

According to his personal story, at the age of 12, while walking through a cornfield with another friend, he had a paranormal experience. He was taken in a UFO vehicle and was transported towards inner earth. Here he lived for 6 months among the residents of the hollow earth. «*You can imagine the wonder of my parents, especially my father who was in the military when I disappeared, to mysteriously return 6 months later. It was because of this experience that I believe my father was certain that I was taken under his wing at the Pentagon and later assigned to serve at Area 51."* And he maintains: "*I am not my father's biological descendant, but an adopted son just like my sister. My sister was killed by what is called 'the secret government'. I was able to fight his negativity with my mind, which is stronger, and I survived his attacks. It is my knowledge through my guide Zora, an inner earth scientist who is 150,000 years old, that my sister and I originate from inner earth and that our true parents live in inner earth. When our father took us up as his foster children we did not speak any language known to any surface culture. I have an unknown blood type. I have never had a disease of any kind. My blood has been medically tested and destroys all viral infections when combined with other blood samples in a laboratory setting."* The Colonel claims that the residents of the Hollow Earth have the ability to split the ocean floor and create a vortex, as shown in the Bermuda Triangle. It says that there are 7 different levels in these vortices, and equipment and beings are brought and placed in correspondence with these various levels. He also holds that vortices act as gateways or exits to the hollow interior of the earth. « *There is more than one triangle in the outskirts of the Florida area, in Lake Erye, and in the outskirts of the coast of Mexico, one in the surroundings of Japan as well as other geographical locations on earth. These are called 'quiet zones'. These portals allow the creatures inside to go outside and come back inside in ways such as the Sasquatch, Loch Ness…etc. All the planets are hollow, as is the sun, which is really a planet. There are civilizations on the sun, which have colonies in the subterranean regions of the Earth."*

« To locate an entrance to the inner earth, wherever you are underground, all you need is your compass. The compass will begin to rotate as if you were standing at the North Pole at the entrance to the tunnel leading to the inner earth. When I left the service, I had no more means to go to the hollow earth. It was necessary to find another way. Together with a group of interested seekers, we chartered a plane that took us to the very edge of the North Pole." According to the Colonel, these people always ask permission to work with nature, they ask permission from plants before consuming or cutting them, they ask permission from Mother Earth before building on it, and they build with the arrangement that best suits them. maintaining harmony with the environment, a practice similar to that of the American Indians; therefore, they seek to preserve a state of harmony always; wanting to be one with nature at all times; they are much more spiritually advanced than the surface dwellers and greatly respect mother earth. He also comments that the atmosphere is crystal clear, in general there are clouds at certain times, but never like rain clouds. The temperature is a constant 73° Fahrenheit (approx. 36° Celsius). The people of the interior speak directly to the animals, and the animals speak directly to the people of the interior. There is no need to accumulate as everything is freely available, no need to create in abundance as everything is sufficient and abundant. An exchange or barter process is more common than trading in money. This is basically a utopian culture without the depression that leads to violence. There are no parties seeking to wage war and gain domination over the others. There is none richer or poorer. The people of the interior are allowed to enter the space of their imagination, if they wish, and from

there they can create. Disease does not enter their bodies - it is not allowed. « *As the humanity of the surface approaches the incoming 4th dimensional phase, the people of the interior of the earth will come closer and work more deeply with us on the surface. People on the surface are currently so involved with the sense of 'I' that they cannot live together in harmony. The people of the surface who try to reach the inhabitants of the inner earth, through meditation will receive them. Children now being born are becoming better able to use the entirety of their brain, which is common practice indoors." "One of the first things we were shown inside was its ability to perform interplanetary travel and time travel. The basis of time travel is related to the curvature or bending of space, which comes with the power of meditation and the acceptance of the self as an unlimited being. If you train your mind on a subconscious level that you are an unlimited being, all things are possible. On the surface, the abilities to experience this infinite power are most easily awakened at portals such as Mount Shasta, which serves as a space-time portal leading directly to inner earth. Once you are in the surroundings of Mount Shasta you are taken to a state of harmony. In my experiences on Mount Shasta the Telosians are projecting an aura of great harmony in a lovely atmosphere."*

It is difficult to mention the intraterrestrial world without referring to UFOs or the many underground military experimental complexes of the armed forces, such as Area 51. According to the Colonel, and agreeing with other military personnel, of everything I saw in Area 51, the 95% remain hidden from the public. Going in is like going to another world, where they are afraid that other countries and other parties are going to get "this" information. His thoughts are: « *If we admit that the earth is hollow, with a central intelligence in it, this will cause discord and fear.* » This fear process is generated by private companies trying to control and advance their own personal needs and agendas (goals) with Area 51. «*I left the air force because of their domineering ways, for those who tried to act like the monsters of the control, who were trying to shut down my ability to think and act creatively, and by accepting their orders to 'not talk about such and such information' - in which they assume one will automatically obey. Due to my outgoing desire to share the information and to inform the general public, my pension service and all my benefits and rights such as the use of dental and medical services were taken away from me. I was with the military for 13.5 years, from the base at the Pentagon and then Area 51. The genetic engineering that is being done is being done with our younger generation. The missing children, whose photos were commonly seen in the markets in the past, were abducted and taken there."* He claims that level 16 is the level of genetic engineering, where our children are being used for experimentation on longevity and the powers of the mind. The biggest force behind this is what is called "the secret government." « *There are civilians from the secret government who are in control in various areas... There is a network of underground tunnels that reach Europe, South America and all the continents. And there is an interconnection of this great network of tunnels across the globe, which many governments use... May God bless you and be with you. Colonel Bill Faye Woodard.* » (Colonel Billie Faye Woodard, telephone communication of January 10, 2002. [Greg Gavin www.Onelight.com - registered 01/10/2002 and transcribed 01/14/2002]. Source: ERKS)

If the Earth is hollow, this implies a greater complexity to justify the natural formation of said sphere. So many theories only try to keep out the creative hand that has really intervened in these matters. One of the most archaic and magnificent legends of primitive humanity is that of a portentously rich, powerful and wise kingdom that exists hidden in the bowels of the Earth. It is said that there reigns a monarch who could be, if he wanted, the King of the World, the Maha Choan, Lord of Civilization and Time. The famous Roman historian Pliny reports that the inhabitants of the marvelous island of Hyperboreas (mythical country of great beauty) managed to escape the cataclysm that sank that Eden under a shroud of ice, through caverns and tunnels that reached to the south of the current Germany. That is to say, 1,200 nautical miles of underground galleries joining the Arctic circle with the temperate lands. This is equivalent to 2,246km. Trying to imagine what such a journey must have been like is already difficult and exhausting. Doing it must have been hellish. A few thousand human beings carrying the victuals they managed to take, taking their children along an endless path of absolute darkness, always fearing sudden abysses, having to choose the correct galleries, fearing getting lost irretrievably. Often found in blind anchovies. A journey of endless months, during which they undoubtedly had to supplement their diet with the blind vermin that have chosen to dwell in those places where the astral light never reaches. Insects, bats, worms, wacky fish. The writer Peter Kolosimo mentions that in the Amazon, an explorer accidentally lost in a vast underground labyrinth went deeper and deeper into it, blind in his anguish. Suddenly, the man found himself in a place where the walls of rock and earth were illuminated "as by an emerald sun." These caverns, illuminated so strongly by that luminosity, stretched out indefinitely. He says that the adventurer also found himself before an enormous insect that looked like a spider of colossal dimensions and also a colossal appetite, for which reason he had to flee as quickly as possible. During his escape, the explorer saw at the end of one of the illuminated galleries some shadows that resembled human beings. The traditions of the Tibetan lamaseries also affirm that there are vast underground spaces in which a source of energy radiation abounds that emits a kind of green luminescence capable of substituting advantageously for solar rays since it stimulates the growth of vegetables and admirably prolongs the life of the sun. human life while retarding the aging of the body and mind.

In the USA there are other curious contributions that reinforce the folkloric traditions about the "emerald sun". A gold prospector, named White, recounts that during one of his forays in search of ore, he entered some natural caverns in 1935. He recounts that he advanced far into the interior of the earth until he suddenly found himself in a plaza or hall vast proportions where lay hundreds of human corpses that seemed naturally mummified. Some appeared seated on carved stone seats; others were stretched out on the smooth, level stone pavement, in strange postures, as if sudden death had frozen them in the midst of dancing movements. White adds that these beings were seen dressed in clothing made of a material similar to leather, although it was clearly something else. Around them glittered great statues of molten gold. We said "glow" as the whole place was clearly illuminated by a strange green fluorescence. White's account caused a stir and an expedition was soon organized in search, without hesitation, of the golden statues. The expedition failed to find the place indicated by the gold prospector, although they ventured through several labyrinthine caverns in which they had to spend a lot of time and precautions to avoid getting lost. Later, the journalists interviewed an old miner who admitted that he also knew those underground places. The journalists subjected him to detailed questions, in order to confront his version with details unknown to the public of White's account. Both accounts coincided in almost every detail, and the differences were not contradictions but simply stemmed from two different perspectives on the same place. The old miner also added that for no reason would he indicate to anyone the entrance to that place, since it was a "cursed town" that could unleash horrible calamities on people if some imprudent intruders broke certain seals. It was obvious that the old man felt a superstitious, invincible terror in relation to those caverns.

Also in the US another impressive story emerged regarding the deep inhabited caverns. Peter Kolosimo himself reports that in 1920 an Indian guide from California, named Thomas Wilson, gave him the story of a strange adventure that would have happened to his grandfather. Wilson says that on one occasion the old Indian entered some narrow canyons and gorges that soon became subterranean galleries. The man, with the inquisitive stoicism typical of his race, continued his exploration, aided by a luminosity that was dim at first but later became more intense. Finally he ended up in a great underground city where he remained for some time living among strange people who welcomed him with discreet charitable hospitality. The old man was not in a position to provide many details about the life of these people, except that they were very strange, spoke an incomprehensible language, and ate a certain kind of food that he did not find very tasty and that was not of natural origin. Perhaps with this the Indian only meant that they were not any kind of food that he was familiar with. The old man also mentioned that those individuals wore clothes made of something that looked like leather but was not leather. That description obviously leads us to think of plastic materials, but at the time both Thomas Wilson's grandfather story and White's emerged, plastics had not yet been developed, let alone a kind of plastic with the one that could make suits.

There are other Indian legends about vast networks of tunnels, some of which go down to such depths that the heat becomes unbearable and it is impossible to breathe. Skilled at using natural caverns as hiding places, the Apaches have stories that we find hard to believe. For example, one of them talks about communication, through deep caverns, between their territory and... the Inca highlands! They say that a group of their ancestors, fleeing before the fierce attack of another tribe, had to take refuge in those tunnels. Once in them, they undertook a journey that took them "several years" and that took them to the distant country of South America. And something more impressive: according to those Indians, it would be very easy tunnels, dug by the hand of intelligent beings. According to the Russian naturalist Ossendowky, Agarthi is the secret heart of wisdom and intelligence, and its origin would go back no less than 600,000 years ago. That is to say, Agarthi would not have been built by our Homo Sapiens species, which in those years barely managed to outline the characteristics of today's man. There are numerous opinions that aim to relate the Agarthi with the ancient pre-human civilization of the disappeared continent of Mu. Unfortunately, those who provide us with details about Agarthi and about Mu do so in a way between fanciful and naive, basing most of the time on sources as dubious as spiritualist revelations or the perception of cosmic voices. Apart from this, they usually add to their descriptions certain supposed "secret teachings" of a spiritual nature that finally end up being the elaborate repetition of "Love one another" that Jesus Christ expressed with such simplicity. However, the myth is powerful and is alive in India, where it is expressed in different versions but which coincide in indicating the belief that in that kingdom there is a superior science and an understanding of reality that demands an ethical sense from its inhabitants. fundamentally different from ours. So much so that the inhabitants of Agarthi have chosen to remain locked up in their underground borders, not allowing more than a select few to pass through, who apparently never return to rejoin the world of common men.

The little that is known about the investigations of the Nazis has come to light through anti-Nazi versions and with the evident intention of ridiculing Hitler. However, the Russians had begun a vast program of scientific research in Tibet, following the course of existing traditions in the Himalayan Buddhist lamaseries, including the study of initiation practices and possible telepathic communication with intelligences from the outer cosmos or from the cosmos. inside. China has not disclosed whether it has continued such investigations after the cooling of diplomatic relations with Moscow and has also refused to allow new Russian expeditions access to those places. Also in their own territory, the Russians have been able to carry out interesting explorations of these enigmatic subterranean civilizations. In Azerbaijan, for example, where several scientific expeditions were sent based on the scandalous superstitious comments about the "Bottomless Pit", a vertical funnel, apparently formed by nature, opens in that region of the Caucasus. People spoke of howlings, blows, noises of forge and shuddering wails ascended and that sometimes a bluish glow illuminated the depths while the walls also seemed to acquire that dim luminosity. When making the descent through the "chimney" of the cavern, they had to give up trying to reach the bottom itself, since its depth was too deep even for modern means of speleological exploration. They then directed their investigations towards the cavities of the contour, in search of some system of tunnels that would allow them to descend gradually without completely losing contact with the central "chimney". Thus they discovered a maze of astonishing caverns. Successive explorations have verified that this cave system reaches enormously distant places in the Caucasian region and in Georgia, towards the South.

The natural inclination to suppose that these caverns had been occupied by prehistoric men was at first confirmed by the discovery of human bones and some crude and easily determined rock inscriptions. However, a more detailed examination revealed that the bones were much older than the artificial excavation of many of the caves and galleries that interconnected the natural tunnels.

Finally the Russian speleologists discovered that there was a whole system of tunnels converging towards certain main arteries excavated in the depths. Unfortunately, the exploration has not yet been exhaustive due to the numerous obstructions caused by landslides. Despite everything, the already explored network of artificial galleries is surprising. Almost all of them lead to circular chambers or plazas, with vaulted ceilings, from which new ducts diverge. Other excavations of special shapes are also noted: empty niches, vertical wells dug plumb, and extremely narrow ducts as if objects of little volume and great weight had slipped through them. The exploration of a comparatively open gallery led, after several kilometers, to an extraordinarily wide square, with a vaulted ceiling more than 20m high, which bore unquestionable traces of having been the work of intelligent work, highly skilled in working on stone. and in the design of vaults that are almost perfect in their original shape, and with absolute mastery of the layout of straight and firm walls that preserve their architectural design despite their incalculable antiquity for now and the telluric movements that they have had to endure.

The Voices of Hell

There is an archaeological movement in Russia that maintains that the system of galleries would extend, with various other exits, to beyond the mountain ranges of the Chinese and Iranian border, and it is also supposed to be connected with the caverns recently discovered near the border. with Afghanistan. In other words, a whole inexplicable labyrinth of tunnels, in which unknown beings carried out a titanic work of sapping to unite remote places with each other, which it would undoubtedly be more comfortable to reach on the surface even if it were on foot and not on the back of a good Bactrian camel. The idea of a Hollow Earth made it possible to more clearly justify many strange phenomena that are still enigmatic for science, such as certain movements of the tides or the notable rises in temperature that have occurred in the stratosphere. It was only the experimental evidence of space flight that made it possible to definitively banish that other form of "Hollow Earth" proposed by the German scientist Herr Horbigger whose greatest merits are oriented towards the astrological interpretations he made and that led him to establish eras of 2,100 years, based on the magical character of the numbers 3 and 7.

«*Russian Scientists Discover Hell!* », This is how an unexpected case in Russia was cited that shocked believers and non-believers, scientists and everyone who received the echo of the news. Something that has aroused curiosity since ancient times is the famous "Hell". According to Greek mythology, there are several hells below the Earth, in the world of the dead under the rule of the god Hades. The possibility that the Earth is hollow, that it can be entered through the North and South Poles, and that secret civilizations flourish within it, has fired the imagination since time immemorial. Thus, the Babylonian hero Gilgamesh visited his ancestor Utnapishtim in the bowels of the Earth; in Greek mythology, Orpheus tries to rescue Eurydice from the underground hell, and Hercules goes in search of Megara; it was said that the pharaohs of Egypt communicated with the nether world, which they accessed through secret tunnels hidden in the pyramids; and the Buddhists believed - and still believe - that millions of people live in Agharta, an underground paradise ruled by the king of the world; For its part, Norse mythology also speaks of heroes who searched for the caves and doors that lead to this place, so much so that the god Odin had a whole palace of heroes killed in combat, called "Valhala"; The Hebrews believe in the Bosom of Abraham, something like the Greek "Elysian Fields" in a world surrounded by darkness under the Earth called Sheol. The famous "Hell Hole" is the story that was broadcast about three episodes of "Trinidad," a network show that aired in the early

1990s. Trinidad also published an article on its Internet mailing list. It is said to have been translated from the original Finnish newspaper "Ammennusastia". The story involves a team of geologists drilling 14.4km into the earth, far south in Siberia into long natural caves, to study the makeup of the earth's crust. They lowered microphones into the hole and were stunned to hear the screams of people suffering in horrible agony. They could only assume that they had reached the mythical Hell.

A newspaper article in Finland added more details: "*A luminous gas was projected from the hole when it was being drilled.*" *Resembling a brilliant diamond, a being with the wings of a bat arose, then the words in Russian were attached to this: "I have conquered", "visibility against the sky".*» In other words, something similar to sunlight emerged from that hole almost 15km deep. The leader of the research "Azzacov", said: «*The deep center of the Earth is hollow!... Temperatures of 1,100ºC were recorded... we could hear thousands, perhaps millions of people in the background, screams of suffering souls." "The information we collected is so amazing that we are sincerely afraid that we could find ourselves there.*" In other words, they were terrified to even think that at some point they could be there, in that place where the screams came from. Half of the scientists reportedly refused further drilling, and the entrance to the chain of caverns was later plugged. Apparently, one of the last events caused them to make a hole and they saw light come out of there and also an animal that they later identified as a pterodactyl —according to a rumor about it.

Perhaps the most moving myth about the Hollow Earth is the one that identifies the Agarthi with Hyperborea. According to many researchers with esoteric inclinations, the superior beings that inhabit there are so perfect that they could be defined as "myths clothed in matter." A brilliant synthesis of this new legend is delivered by the writer Miguel Serrano in his work "The Golden

Cord", although the theme radiates and penetrates most of his mature work. According to this legend, the Earth would be hollow inside; At a depth of 800 km from the surface, a vast "inner world" would open up, with respect to which gravity would act in the direction of the crust, since the main planetary mass was found there. In the center of the planet, like a radiant heart, there would be a fiery and luminous nucleus that would provide light and heat to that "inner" world. This legend holds that within this globe, upside down in relation to us, there are tense lands, seas, mountain ranges and rivers, a whole small planet involved by our own, with less gravity that would allow its inhabitants and vegetation to reach enormous statures. In addition, the virtues of the radiant heart of the Earth would favor life much more than sunlight and would rejuvenate those who managed to enter it at an advanced age. At both poles, this legend says, the main entrances to the Inner Planet Earth are hidden, although there would also be other access routes in the Himalayas, the Andes and certain mysterious caverns that are lost in the deep darkness of Mother Earth. But how did all this come about? Who, when and how created, or created, this planet with these galleries, internal oceans, abysmal holes, beautiful kingdoms and kingdoms of darkness? And the demonic beings that are said to also dwell there, the dead and even magical beings, what about all this? Where did it come from?

The Abode of the Dead

The Spanish translation of the Book of Mormon indicates, in the 2nd Nephi text (chap. 24:19), that Satan -it is understood that he is spoken of- will be thrown into the Abyss, into the hole of the Earth, where there are rough stones and impassable walls. This definition, similar to those of Isaiah and Ezekiel, mentions it like this: «... *but you are thrown out of your grave like an abominable branch, like the residue of those who were killed, pierced by the sword, who descend to the stones of the abyss; like a carcass trodden underfoot."* So even though it's a gigantic hole, it has boundaries that are set by hard rock. These descriptions can correspond and fit with the many that Enoch gave, among them the part in which he is taken where there are beings of fire and the weapons of the angels: «They *took me to the house of the storm, on a mountain whose top touched the Heaven, and I saw the mansions of the luminaries and the treasures of the stars and thunder, at the ends of the abyss where are the bow of fire, its arrows and quiver, the sword of fire and all the lightning. Then they took me to the waters of life and to the fire in the west, the one that collected all the sunsets. I came to a river of fire whose flames run like water and empties into the great sea that is on the west side; I saw great rivers and I came to great darkness and to where no carnal being walks; I saw the mountains of winter darkness and the place where all the waters of the deep flow; and I saw the mouth of all the rivers of the earth and the mouth of the deep."* (1 Enoch 17:2-8) The Sumerian poem "Ishtar's Descent into the Underworld" speaks of a land of no return, of darkness and misery, where prisoners eat mud for bread and drink muddy water for beer -he adds that there some Anunnaki live.

In his fabulous travels, carried by the Lord, Baruc recounted that he was taken into the interior of the Earth and observed Hades. His companion told him: «*And this is the underworld, which in turn is also similar to it, in the sense that it also drinks a cubit from the sea, which does not sink at all. Baruch said: And how (does this happen)? And the angel said: Listen, the Lord God made 360 rivers, of which the chief of all are Alfias, Abyrus and Gericus, and because of these the sea does not sink.*» (Apocalypse of Baruch or 3rd Baruch 4:6-8) This seems to say that there is no flood because the water of the oceans is carried in flow by branches, rivers, everywhere. Down below there would be no sinking or flooding either because of these well organized and situated rivers and limitations that have been placed on the seas, showing how perfectly everything on this planet was laid out and terraformed in a slow and wise manner. Jesus once recounted: « *There was a rich man, who dressed in purple and fine linen, and made a splendid feast every day. There was also a beggar named Lazarus, who was lying at his door, full of sores, and longed to be satisfied with the crumbs that fell from the rich man's table; and even the dogs came and licked his sores. It happened that the beggar died, and was carried by the angels* **to Abraham's bosom**; *and the rich man also died, and was buried. And in Hades he lifted up his eyes, being in torments, and saw Abraham afar off, and Lazarus in his bosom. Then he, crying out, said: Father Abraham, have mercy on me, and send Lazarus to dip the tip of his finger in water, and* **cool my tongue**; *because* **I am tormented** *in this flame. But Abraham said to him: Son, remember that you received your goods in your life, and Lazarus also evils; but now this one is comforted here, and you are tormented. In addition to all this, a great chasm is placed between us and you, so that* **those who want to go from here to you cannot go here, nor from there**. *Then he said to him: I beg you then, father, that you send him to my father's house, for I have five brothers, so that he may testify to them, so that they too may not come to* **this place of torment**. *And Abraham said to him: They have Moses and the prophets; listen to them He then said: No, father Abraham; but if someone goes to them from the dead, they will repent. But Abraham said to him, if they do not listen to Moses and the prophets,* **they will not be convinced even if someone rises from the dead**." (Luke 16:19-31)

Death is a topic that is not immune to thousands of discussions and uncertainties, " *nothing is certain except death* ", they say out there, and the location for those spirits that are no longer in the body is, in the past, visualized in the bowels of our globe. Jacob assumed that when he died he would go to the sides of the abyss, specifically to the place called by the Hebrews "Sheol" (usually transcribed as "sheol"). That Sheol was called "Hades" by the

Greeks, and the idea was promoted that the whole region was "Hell". But, was Joseph, son of Jacob, a young man who did not please God? The scripture says that it did please the Lord and therefore did not deserve Hell. Why did his father say that he would go to Sheol together with his son? Why didn't he say that he would go up to heaven with him? Well, it says literally: « *Then Jacob tore his clothes, put rough clothes on his waist and mourned for his son for many days. All his sons and all his daughters rose to comfort him, but he did not want to be comforted, saying: "I will go down in mourning with my son to Sheol!" And his father mourned him."* (Genesis 37:34-35). Jacob was convinced that when he died he would go to Sheol and not to heaven: « *But Jacob replied:—My son will not descend with you, for his brother is dead and he is left alone; If any disaster befalls you along the way you are going, you will bring my gray hairs down in pain to Sheol."* (Genesis 42:38, 44:29 and 44:31). And in this way Job himself also speaks of himself, who was a just man in the eyes of God, desiring the descent to Sheol (other examples in Ezekiel 14:14 and 14:20). The Bible says: "*Yehovah gives death and life; He brings down Sheol and returns."* (1 Samuel 2:6) Because we all must return from below. We see this when it speaks of going down "in peace" to Sheol! (Job 21:13). Although when referring to "below" one thinks of hell, not everything is so simple. When dying, not all spirits are in the same place (1 Kings 2:5-6 and 2,9, Psalms 89:48 and 141:7, Ecclesiastes 9:10, 1 Samuel 28:14, Hosea 13:14 and Jonah 2:2). The Apostle Philip wrote: " *You are either in this world or in the resurrection or somewhere in between. Please God they don't find me in these! In this world there are good things and bad things: the good things are not the good and the bad are not the bad. But there is something bad after this world that is really bad and that they call the "Between", that is, death. <u>While we are in this world</u>* **<u>it is convenient that we strive for the resurrection</u>** <u>*so that* — **once we put off the flesh** — *we find*</u>

ourselves at rest and do not have to wander in the "Intermediate." Many actually go astray. It is therefore expedient to leave the world before man has sinned." (Gospel of Philip 1:63)

Samuel's own story is very clear. The prophet had already died and King Saul sought a summoner of the dead to call Samuel from the dead: "*And the woman said to him, Behold, you know what Saul has done, how he has cut off the earth from the dead." evocative and diviners. Why, then, do you put a stumbling block in my life, to make me die? Then Saul swore to him by the Lord, saying, As the Lord lives, no harm will come to you for this. The woman then said: Who shall I make you come? And he answered: Make me come to Samuel. And when the woman saw Samuel, she cried out with a loud voice, and the woman spoke to Saul, saying, Why have you deceived me? for you are Saul. And the king said to him: Do not be afraid. What have you seen? And the woman answered Saul: I have seen gods coming up from the earth. He said to him: What is its form? And she answered: An old man comes, covered with a cloak. Saul then understood that it was Samuel, and humiliating his face to the ground, he made a great reverence. And Samuel said to Saul: Why have you disturbed me by making me come?*» (1 Samuel 28:9-15) The first great prophet of Israel, a righteous man in the eyes of God, did not come down from heaven but "came up" from under the Earth. From the beginning of creation, either spoke of the underworld. Both the Mesopotamian tablets, the scrolls of Enoch or the writings of Moses, state that the early Earth was divided into 3 sections: top, middle and bottom. Above was Heaven, in the middle was the planetary surface and thus our abode, and below were: the prisons of the fallen angels; The world of the dead; the climate control systems, the seasons, the clouds, the rain, the sea and in general all the functioning of the planet Earth; and the bases and cities of angels. The prison section was called Tartarus, the world of the dead was called Sheol/Hades, and the rest of the angelic lands would come to be called in different ways, including Agartha/Shambhala by the Tibetans; although all that region, in general, would always be called The Abyss (Tehom). The Hebrews nicknamed this site more explicitly "Abadon" (Destruction/Slavery) to differentiate it from Sheol; while the Greeks called him Apollyon, to differentiate him from Hades, the abode of the deceased.

No people have been exempt from having their point of view and belief in this regard, as in the case of Taoism, where it is believed that the underworld is ruled by 5 legendary judges (as in the Gnostic perspective), called: Huang-Chuan, Chu Jiang, Song-Di, Wen-Shu and Yan-Lo (also called Wang, which is god and judge of death, possibly equivalent to the biblical Belial). That place under the Earth receives the name of Huang-Chuan, to which the souls, "Yin", return after death. The definition of the spirit "yin" is the same as that passed to the Arabic, "Ginn", and later to the Spanish "genius". In the region of Sheol/Hades, according to the prophet Enoch, there were 4 divisions, one of them would be called by Jesus "The Bosom of Abraham " (see reference in Luke 16:20), and by the Hellenes "The Elysian Fields".

The Hebrew historical references in this regard are very extensive, as are the descriptions of the past throughout the world, from the so-called "region of the Abzu" (South Africa or the Underworld) by the Sumerians, passing through the Lower World of the Egyptians. or the "kingdom of the Emerald Sun", called like this by other peoples of America. Jesus himself had to go down to preach to the imprisoned angels (1 Peter 3:18-20) and bear witness to the light in Hades (Ephesians 4:8-10). Jesus, who died and rose again, testified that no one ascends to heaven if he has not risen and that by achieving victory over death he took possession of that kingdom of darkness: "And what is it that he ascended, but *that had also first descended to the lowest parts of the earth? The one who descended is the same one who also ascended above all the heavens to fill everything."* (Ephesians 4:9-10) So what about the legend of Hell? They are outdated. The Catholic Church potentially promoted this idea, which then greatly absorbed Protestantism and even took root in Islam, as they instilled fear in people. The versions of the Bible that spoke of Gehenna (name of the "Lake of Fire") were translated into the Latin voice "Inforno" (in heat) and mixed with Greek legends about places of torment in Hades. The true hell or Gehenna is the "Lake of Fire" (the word "Lake" translates into Greek: "Limni", from where the voice comes: "Limbo") that will be inaugurated by the Son of Perdition, and therefore will be the fate of those deemed to be brought there after the day of the Great Judgment: *"And I saw a great white throne and him that sat on it, from whose face the earth and the heaven fled away, and there was no place found." for them. And I saw the dead, great and small, standing before God; and the books were opened, and another book was opened, which is the book of life; and the dead were judged by the things that were written in the books, according to their works. And the sea gave up the dead that were in it; and death and Hades gave up the dead that were in them; and each was judged according to his works."* (Revelation 20:11-13). For this reason it is reiterated: *"But the cowards and unbelievers, the abominable and murderers, the fornicators and sorcerers, the idolaters and all liars will have their part in the lake that burns with fire and brimstone, which is the second death."* (Revelation 21:8)

Hades is the "intermediate" point before the Resurrection of the dead, but it is an area of the Abyss, and the Abyss itself -where ancient spirits of darkness dwell- is part of the world under the Earth, where there are also dungeons like jails for rebels against God, also galleries and domes where dark beings live and also a paradisiacal world known as Agarti. Hermes said: *"Now, the pit of the world is a roundness in the form of a sphere, totally invisible because of this quality or shape, and so it is that, the higher you climb into it to look down, from there you cannot you will be able to see its bottom, where many think it is like space. We say that it is visible only because of the sensible figures whose images we see inscribed on it, in the manner of a painted picture. But the truth is that the Sphere is always invisible in itself, from which its bottom or part, if a sphere has a bottom, the Greeks call Hades, from the Greek "idein" which means "to see", because it cannot be seen the bottom of a sphere. Whereby sensitive forms are also called "ideas" because they are visible concepts. Due to the fact that they cannot be seen, because they are at the bottom of the sphere, the Greeks called Hades what we Hell.»* (From Hermes Trismegistus addressed to Asclepius, verse 17. Corpus Hermeticum) Later he teaches him: «*-Listen, Asclepio. When the soul is separated from the body, it passes under the authority of the Supreme Dáimon for examination of its merits, and, if from careful scrutiny it appears pious and just, he authorizes it to dwell in the place that corresponds to it, but if it sees it dirty with traces of crimes and stained with vices, precipitates it from above to the depths and surrenders it to the storms and whirlwinds, always found, of air, fire and water, so that, dwelling between Heaven and Earth, it may be permanently dragged by the mundane waves and agitated between endless sorrows, because even eternity opposes it, because it is subjected by an imperishable sentence to an endless torture. So be aware of what you have to be ashamed of, fear and beware of, so as not to fall into the same thing. Because the unbelievers, having committed the crime, will be forced to believe, not with words but with deeds, not with threats but with the very suffering of the punishment.*" (Verse 28)

II.

THE KINGDOM OF THE FATHER

THE UNIVERSAL FATHER

"Science is the aesthetics of intelligence."
Gaston Bachelard (1884-1962)

To have a notion of how everything began from its darkest and most primitive beginning, we cannot, obviously, refer to archeology but to "divine inspiration", since every recorded mention speaks of the primitivism of the gods before the visible. But what is divine inspiration? Men cannot by ourselves, even if we wanted to, get to know the mysteries of the universe without "superior" help. If there are humans in multiple forms in the cosmos, who do not age, who do not get sick, who are therefore basically eternal, they are the ones who could contribute more to us on this mysterious and unknown path, even though they also have higher realities above them. However, in recent decades it has been claimed to have communication with them, beings from space, but it has been observed that many of these have posed as God, as some already did in the millenary time, saying only what they want the spacemen to say. who listen to them are created. So, if we want to know further back, the beginnings of the creation of everything that exists, it would be easier if the Creator himself told us what happened, or people close to Him. Thank God -pun intended- He has advised men, both from heaven and from Earth in the mysteries of the multiverse in which we coexist with beings from other worlds and levels, that we are not capable of understanding in our dawn here, in this immeasurable "field".

Although, we can now find new ideas such as the TOE (Theory of Everything = Theory of Everything), which incorporates studies of elementary physics, metaphysics, quantum physics, and other research on psychology and parapsychology, analyzing the Unified Field. All this leads us to the same thing: the immaterial universe. And if in fact this is so, the theories about the origin of Creation take us in the same direction, and the beginning of this path would be located in the Bosom of the Father, the home from which all things have emerged. It was written many centuries ago: «... *Or will the world return to its nature before-time, and return to the silence of the primeval age? ...*» (2nd Baruch 3:6) This scribe (son of Neriyiah, son of Maaseiah, son of Zedekiah, son of Asadiah, son of Hilkiah), dating from the time of the deportation of the Jews to Babylon, speaks of the time when time was not considered. There was, however, a world or universe before the "time", or framework that began after the Big Bang, in the era of "silence", which gave rise to this universe subject to the law of space-time. We can also find what was written by the prophet Baruch about an invisible universe: «*You show great acts to those who do not know; That breaks the box of those who are ignorant, And the lightest is dark, And reveals what is hidden from pure matter ...*" (2nd Baruch 54:5). That "invisible" place appears to be the root of the visible, but to investigate it we have to understand who its maker is and how and why he determined to establish observable bases millions of years ago to the human eye.

The apostles had many experiences with the risen Christ, which helped them broaden their understanding, especially about existence, the origin of things, and their subsequent destiny. In one of several revelations, the Lord speaks to them about the essence of the Universal Father, from whom proceed all things subject to his will, and about how those who think they are wise seek to understand the universal mechanism and life: "...the Savior *answered: "Keep it in mind: all men born on Earth from the creation of the world until now are dust. They are looking for God, what he is, what he looks like, and they have not found him. However, the wisest among them have lectured on Him, basing themselves on the ordering of the world and its movements. But his lucubrations did not reach the truth. Well, in three ways the philosophers would explain this arrangement: hence they did not agree among themselves. Some say that the world moves by itself. Others, that it moves by providence. Some more, than by fate. Well, everyone is wrong. No, none of these hypotheses is close to the truth, since they are human judgments. But I who have come from the infinite Light, I do know. I am the one who knows: that is why I can speak to you about the exact nature of truth. Because everything that comes from himself is corrupted life, made of himself. Providence does not contain in itself wisdom. And the inevitable cannot be discerned."* (Sheneset Gnostic 1:2) These words had been spoken by Jesus just before he was elevated to heaven. Looking at the biblical gospels we find the words of the apostle John referring to the beginning of everything: «*In the beginning was the Word, and the Word was with God, and the Word was God. This was in the beginning with God. All things were made by her, and without him nothing was made that was made. In it (the Word) was life, and life was the light of men. The light shines in the darkness, and the darkness did not prevail against it.*" (John 1:1-5)

Some associate that "Word" with the Son of God, but there would not be much contradiction in this or differences relevant to the case. The issue is that this Word or Son was the "light" in what Baruc calls "before time." That word was the universal Law, the first to emerge. This is what they say happened: in the beginning there was only the Father, then the Son, understood as the development of the first thought of the Father and his company. Subsequently, more thoughts of the Father were revealed, consistent with his co-workers, who were with him in a light that issued from the Father, they were him and part of him, though each an independent entity or life form. There is, therefore, no sense of division in the Father, since everything is ONE in Him and with Him. « *Glory of all things is God, and his divine being, and his divine nature. The principle of all beings is God, and theirs is intelligence, nature and matter, wisdom that shows what all things and each one are. Principle is the divine, and it is nature, energy, necessity, end and renewal.*» (Hermes Trismegistus a Tat, Corpus Hermeticum. Treatise XVII, verse 1. Original incomplete and without title) Elsewhere, Hermes Trismegistus says to Asclepius: «The *other name of God is that of "the Father", now because created all things: the father is the one who creates."* (Chap. XVI, verse 17) And on another occasion he also teaches Asclepius: «- *God is not intelligence, but the cause of intelligence existing. It is not spirit but cause of the spirit's existence. It is not light, but the cause of the existence of light. Where God must be worshiped with those two names, which only belong to Him and no other. Because none of the rest that are called gods, nor any of the men nor any demon can in any way be Good, but only God, who is only Good and is nothing else. All other beings are incapable of containing the nature of the Good: they are body and soul, and they have no place that can contain the Good. The greatness of the Good is as great as the reality of all beings, corporeal and incorporeal, sensible and intelligible. Behold the Good, behold the God. Do not call anyone or anything good, because it is impious, and do not give God any other name than the only one of Good, the opposite is also impious."* (From Hermes to Tat, universal discourse, lost treatise. Corpus Hermeticum 1 and 2, verse 14-15. Untitled Treatise)

But if in the beginning it was only the Father and his Word, a question that always appears in these cases resonates with us: who is the Father and where did he come from? Jesus answered this to his disciple John, shortly after the Resurrection of the Lord: « *When I asked him if he could understand this, he told me: "The One is a sovereign who has nothing over him. It is God the Father of all, the*

Invisible One who is above all, who is imperishable, who is pure light that no eye can see. It is the invisible Spirit. One should not regard him as a god, or the same as a god, For he is greater than a god, for he has nothing over him and no lord over him. It does not exist within anything that is inferior to it, since everything exists only within it. It is eternal, since it does not need anything. Because it is absolutely complete: it has never lacked anything to be complete. But it has always been absolutely complete in light. It is illimitable, since there is nothing to limit it. It is unfathomable, since there is nothing before it to fathom it. It is immeasurable, since there was nothing before him to measure him. It is unobservable, since nothing has observed it. It is eternal, and it exists eternally. It is inexpressible, since nothing could understand it to express it. It is unnamable, since there is nothing before it that gives it a name. "It is the immeasurable, pure, holy, brilliant light. It is inexpressible, and it is perfect in its immortality. It is not that it is part of perfection, or bliss, or divinity; it is much bigger. It is neither corporeal nor incorporeal. It is not big and not small, it is impossible to say: "How much is it?" or "What kind is it?" for no one can understand it."

«It is not one among many things that exist: it is much bigger. Not that it's really bigger. But as it is in itself, it is not a part of the worlds nor of time, because whatever is part of a world was once produced by something else. No time was assigned to him, since he does not receive anything from anyone. That would be a loan. The one that exists first does not need anything from one that is later. On the contrary, the latter looks up at the former in its light. "Because the perfect is majestic; it is pure and immeasurably great. It is the World that gives a world, Life that gives life, the blessed that gives bliss, Knowledge that gives knowledge, the Good that gives goodness, Mercy that gives mercy and redemption, Grace that gives grace. Not that it's really like that. Rather, it gives immeasurable and incomprehensible light." "What should I tell you about him? His eternal kingdom is

imperishable: it is calm, it is silent, it is at rest, and it is before everything. He is the head of all the worlds, and sustains them through his goodness." "However, we would not know..., we would not understand what is immeasurable, if it were not for one who has come from the Father and has told us these things." "Because the Perfect One contemplates himself in the light that surrounds him. This is the spring of the water of life that produces all the worlds of all kinds. The Perfect contemplates his image, sees it in the spring of the spirit and falls in love with the luminous water. This is the spring of pure water, which surrounds the Perfect One." (Secret Book of John 2:1 to 3:2) Quite well explained, although certainly difficult to understand for the common man accustomed to answers subject to materialism, to definable, pigeonholing concepts, marked in a concrete image and a perceptible concept.

After resurrecting, and before definitively saying goodbye to his disciples, Jesus spoke about the Father to all those closest to him: « *Matthew answered him: "Lord, no one has access to the truth, except through you. So show us the truth." And the Lord answered him: "Ineffable is who he is. No sovereignty reigns over him, no authority, to no dominion is he subjected, nor to any creature from the creation of the world until now, except himself and each one of those to whom he sees fit to make a revelation through him who proceeds. of the first light I am, from now on, the great Savior. Because he is immortal and eternal. Eternal because he has not had a birth: for whoever has one will perish. He was not begotten, for he had no beginning: he who has a beginning also has an end. No one has authority over him, since he has no name: for he who has a name is the creature of another. It could not, therefore, be named. It does not have a human form, because whoever has a human form is the creature of another. And only to himself does he resemble – not to anything, whatever it may be, that you have ever seen or perceived – a strange resemblance he has, far superior to everything and superior to the universe. He looks around and sees himself from himself. Since it has no limits, it is elusive. Imperishable, since it has no peer. Immutable. Infallible. Eternal. Blessed. Since it is not known, it always knows. It is immense. Indescribable. Perfect, because it has no defects. Blessed for all eternity. And he is called the father of the universe." Philip then said: "How then, Lord, was it revealed to the perfect?" And the perfect Savior answered him: "Before anything that appeared was revealed, greatness and authority were already in Him, since He encompasses the totality of everything, while nothing encompasses Him. spirit. He is thought and thinking, reflection, reason and power. Powers all equivalent. Powers that are the source of the whole. And whose prolongation, from beginning to end, was already in his presence, that of the uncreated and limitless Father."* (Sheneset Gnostic 1:4-5).

It is strange to try to reconcile this image that Jesus presents with that of the god of the Hebrews and the events and words that are supposedly recorded about him in the TANAK. They seem to be two totally different people. In fact, analyzing the concepts of God, which Jesus refers to, and those of God, but seen with the eyes of the Old Testament, the same deity is not seen. The Israelites were "Jehovah's witnesses", observing and experiencing his great wonders from Egypt, but, even with all this, Jesus told his compatriots: «*If I glorify myself, my glory is nothing; my Father is the one who glorifies me, the one you say is your God. But <u>you do not know him</u>; but I know him, and if I said that I do not know him, I would be a liar like you; but I know him, and I keep his word.*" (John 8:54-55) Did they not know him? Israel was precisely the one who represented this God, of whom other nations heard and trembled. However, John insisted that "*no one has ever seen God; the only begotten Son, who is in the bosom of the Father, he has made him known.*" (John 1:18) If no one has seen God, whom did Adam, Enoch, Abraham, Daniel, Moses, Isaiah, and many other biblical characters observe? It is not strange that neither Jesus nor his apostles mentioned Jehovah, nor spoke of him, but instead presented a God that Israel did not know, whom the followers of Christ identified as "Aba" (Father). This leads us to turn to the words of Jesus when teaching: «*When you pray, say: <u>Our Father who art in heaven</u>, **hallowed be your name**. Thy kingdom come. Thy will be done, as in heaven, so also on earth.*" (Luke 11:2) If the Father is "in heaven", then he is not Jehovah, since this god went down to Sinai with Israel, accompanying Moses, and even ordered the construction of a tabernacle and then a Temple. At the same time, Jesus says that his "*will*" is not done – with which he cannot be God nor has he come down yet, since he makes it quite clear that the will of "another" reigns here. In addition to this, Jesus speaks of his "*name*" being "*hallowed*", but for what reason? Who has so damaged his name that he needs to be cleansed? In other words, how is it that God's reputation has been damaged? This pushes us to see things differently: «*Not that anyone has seen the Father, except the one who came from God; he has seen the Father.*» (John 6:46).

We have to consider the fact that the deity presented to the Israelites represented God, but was not the Father. For that reason he killed, he got angry, he acted relentlessly, he was jealous, he often did not forgive human errors, he regretted, etc. Qualities that have nothing to do with God, but the Father, who is identified as the only true God, «*the only one who has immortality, who dwells in inaccessible light; whom none of men have seen or can see, to whom be honor and everlasting empire.* » (1 Timothy 6:16). So the Father is everything, is in everything and exists before everything, but He has nothing to do with evil and has not yet been manifested, except in what Jesus made known about Him. Then the Father dwells in each one of us, he is around in everything we see, but at the same time he has his dwelling in the "heights" - although whoever can represent him from the elevated parts is an image of him, but not himself. If someone has come to manifest the Father, it has been the Son, and if someone is to come as a physical presentation of God later on, it will be the Ancient of Days, whom Daniel and Enoch saw -as we have already detailed in our book "Reaching the the Deity". Jesus described it this way: « *The Master of the universe is not only called Father, but the primordial Father. He is the origin of what must be revealed. He is the primordial Father, who had no beginning and who sees himself in himself as in a mirror. He has manifested himself as equal to himself. But this similarity has been expressed as God the Father through himself, confronting those who were confronted, he who was above all the uncreated Father. For the same age as light, which exists before being seen, but whose power does not equal its own. But afterwards there appeared a multitude of finite things, begotten, all equal in age and power, constantly glorified. Their species is called the generation without a kingdom, and it is through it that you yourselves have manifested yourselves among men. And all this multitude over which there is no kingdom is called that of the children of the uncreated Father, God, Savior, Son of God: They are like you. But he is unknowable, he is full of imperishable glory and ineffable joy. Everyone rests in it, everyone rejoices in the ineffable joy of its unchanging splendor and immense joy. Never until now has such a message been heard or known among the aeons and their worlds."* (Sheneset Gnostic 1:9).

Although the Israelites had a specific name for God, the Father has no name: " *We bless you, the non-being, the existence, which is before existences, in the first place there is the one who is before beings, the father of divinity and life, creator of mind, giver of what gives good [things], of bliss! [...] Because there is no one who is active in front of you. You are a unique and living spirit. And you know one, so everything that manifests everywhere belongs to you. Nothing is capable of expressing you. But your light shines on us. [...] You are one, you are a living spirit. How are we going to give you a name? We don't have it, Because you are the existence of all of them. You are the life of all of them. You are the mind of all of them. Because you are the one in whom everyone rejoices."* (Stela 3. The 3 Stelae of Set). So, the one who is before everything, has no name, is not known and is before the visible. And where is his kingdom? Does He dwell in any particular place? Well, we know that " *the Father is in innocence* " and that also "*the perfect Majesty is at rest*" (2nd Treatise on the Great Seth), so the Shabbat identifies the dwelling and state of the universal Father. It is not surprising that they were taught to keep the Sabbath day (all this about the Father, and much more, is detailed in the Gospel of the Egyptians or the Sacred Book of the great Seth). And it is from the Father that the first thing to appear came from: the Word. This manifested itself at the very moment that the Father generated a Thought, so this word is the action itself that is the consequence of the Father's thought. This is how pure light also came into being, by the mere idea of the universal Father, who contains everything in himself and whose power cannot be explained or imagined.

The ancient traditions, despite the obscurity in knowledge about the Most High, seem to have had glimpses of references to Him: « *In the first, three units dwell. Others different from these cannot exist. These are the balance, source of creation: one God, one Truth, one point of freedom. Three come from the three of balance: all life, all good, all power. There are three qualities of God in his home of Light: Infinite Power, Infinite Wisdom, Infinite Love. [...] Three are the inevitable things that God does: Manifest power, wisdom and love. Three are the powers that create all things: Divine Love possessing perfect knowledge, Divine Wisdom knowing all possible means, Divine Power possessed by the union will of Divine Love and Wisdom. There are three circles (states) of existence: The circle of Light where nothing but God dwells, and only God can cross it; the circle of Chaos where all things by nature arise from death; the circle of consciousness where all things flourish from life. [...] There are three obstacles: lack of effort to obtain knowledge; detachment from God; attachment to evil In man, all three are manifested. Three are the Kingdoms of inner power."* (Table 14. Emerald Tablets of Thoth) Specifically in Hinduism, God seems to be reflected in Vedas texts as Brahma, Brahama or Brahmán, who also appears in one of the hymns where he is called Brahmanaspati, or equally as Brihaspati, as the one who « *destroys the Asuras* », referring to the demons. This is reminiscent of the Battles of Jehovah. He also says of him: " *by the power of the sacred word, he triumphed over Bala, dispersed the darkness and made the light shine."* The title "Bala" -or "Vala"- could be speculated to be the root from which the name "Baal" (BA-LA or BA-AL) was derived, since originally the names could be read from left to right or from right to the left, as explained in the cabalistic system of the Temurah. The most striking thing is when he says of Him that he is the *"sovereign owner of prayer"*, as *"priest of the family of the gods "*, and that *he "helps men through virtuous instructions."* Likewise, he affirms that he is " *departed from the truth"* and on some occasions he is invoked as *"the most divine of the gods"* and the *"father of the gods"*, the *"regulator of this world"* and *"supreme owner of all beings"."* These epithets or flattery seem to describe the Jehovah of the TANAQ (Old Testament), since even the affirmation of: "god of gods" is found in a psalm of Asaf: « *God is in the gathering of the gods; In the midst of the gods he judges."* (Psalm 82:1. RVA 1960).

In Arab mythology they defined the supreme and creator god as Allah-Taala or Allah-Toala, considering him the creator of Heaven and Earth (his name was misunderstood by the Greeks, who thought it referred to the gods Orotal and Alilat). There is also the epithet of Allah-Ta-Alai as a very high god, while the others were Al-Jlahat or minor gods. These would be God's company and would be subject to Him in everything, which is reminiscent of the term Benan-Hascha, which means "God's companions." Another identification appears with the pseudonym of Aláh or Al Iláh, who is shown with the crescent and would represent nature, being in turn the father of 3 important goddesses (it is speculated that the epithet of Allah comes from Elah). However, in relation to titles and appellations for the greatest god of all, the Creator, immaterial universal and eternal Father, there are other names in legendary myths. Some seem to lead us to remember Enoch's words where he said that "men would err their ways understanding the stars as gods", but this same translation could not be correct, when referring to "Elohim", that is, it could also be focused on that men would confuse star beings with God himself. Hence Ra in Egypt, Brahama in India, Zeus in Greece, Odin in Scandinavia or Achamán in Tenerife (one of the 7 Canary Islands, in Spain), for example. Achamán (Achuhuran, Achahucanac (Great and Sublime God), Achguayaxerax, Achoron, Achaman (the Sustainer of Heaven and Earth) Abora or Alcorac), was -without its exact pronunciation being known-, the name received by one of the divine mythological entities in which the ancient inhabitants of the island of Tenerife, the Guanches, believed. Achamán held the title of the sky god and was considered the supreme god. Its name literally means "the heavens", alluding to the celestial vault (heaven). At first, according to the Guanches, it was Achamán, a powerful and eternal god that was self-sufficient. Before him there was only nothingness and emptiness, the sea did not reflect the sky and the light still lacked colours. The creatures owed their existence to him, since he created the earth and the water, the fire and the air, and all the life that was in them. Achamán inhabited the heights and sometimes the summits of the mountains to rejoice contemplating what was enlivened before his eyes.

The words of the Israelites and their ancestors are not the only ones that keep mysteries, not even the cultures to which the origin of civilization is attributed. As Plato and many others before and after him made clear, there were ancient peoples of which there is almost no mention today. Even they left references in association with the theme of the God who is above all. Some 36,000-year-old tablets found in the Yucatan reflect the knowledge of these people about celestial aspects that subsequently vanished over time. The emerald tablets were attributed to an Atlantean priest named Thoth and the wisdom he had attained. In one due to his trans-corporal experiences, Thoth had a sumptuous vision: " *Given the key to ascend to the Presence of Light in the Great World. I stood before the Holy One enthroned in the Flower of Fire. He was covered by the lightning of darkness, also my Soul for Glory has been shattered. Before the feet of his diamond-like Throne four rivers of flame ran from his stool, they ran through the channels of clouds to the world of Man. The hall was filled with Spirits of Heaven. Wonder of wonders was the palace full of stars. Above the sky, like a rainbow of Fire and Light of the Sun, the Spirits were Formed. They sang the glories of the Sacred. After the midst of the Fire came a voice: Behold the Glory of the First Cause. I contemplated that Light, far above all darkness, reflected in my own being. I reached, as it was, the God of all Gods, the Sun Spirit, the Sovereign of the Sun spheres. There is One, Even the First, who has no beginning, no end; who has made all things, who governs everything, who is good, who is just, who illuminates, who sustains. Then from the throne, a great brilliance poured down, surrounding and lifting my soul with its power. Quickly I moved through the spaces of Heaven, the mystery of mysteries was shown to me, the Secret heart of the cosmos was shown to me.*» (Table 14. Emerald Tablets of Thoth)

Now, if we want to investigate deeply, we can look for the reference to the One, the Universal Father, the Creator and, ultimately, post-Flood monotheism in Mazdeism or Zoroastrianism. Referring to the teachings of the Persian prophet Zoroaster, we observe the relevance he gives to Ahura Mazda

(Ahura Mazdā) or Ormuz, a name in the Avestan language for a divinity exalted by him as the uncreated Creator, that is, the supreme deity of Zoroastrianism. In the Avesta, " *Ahura Mazda is the highest object of worship* ", the first divinity and the most frequently mentioned in the Yasna liturgy. In Zoroastrian cosmogony and tradition, all minor divinities are also creations of Mazda (read Bundahishn III), somewhat akin to the Hindu idea of Brahama. He is understood as the god of heaven, omniscient and celestial priest, leader of the Amesha Spenta (Zoroastrian divinities, created by Ahura Mazda to help govern creation) or ahuras. According to the Persian prophet, Ahura Mazda is an abstract and transcendent god, without a concrete image, which is why he cannot be represented. The Zorastrians use the term "bind" to refer to the fire and light that are manifestations of Ahura Mazda. The definition of "Ahura" means, literally, the "High Being" -it is masculine-, and "Mazda" means "wisdom", -feminine name. Ahura Mazda is understood through the Good Mind. But this communion, this exchange between Divinity and humans who reach out in search of Him, is unique through the Good Mind. It is communicated and manifested to mortals through their own ethical attributes that are a part of their being. Men and women can choose to be like the Achaa (Righteous) or like Vohu Manah (with a Good and benevolent Mind), or any of the other essences of the very being of God. It is said that He gives His own attributes to mankind in order to progress and evolve towards wholeness.

The Word

So that in this, the Universal Father began to develop his Thought and contemplate it in front of him. As in many mythologies, the first virtues of the cosmos are represented with personified figures, likewise, in this case, the Creator's virtues were manifested around him, beginning with the female image or "The Mother", but all this is prior to what any mythology in the world could consider, since it has only been made known by Christ. What comes out of the mythologies are the teachings of angels to men, but many of the angels did not know the Imperishable Kingdom from where Christ came, so they began to relate the origins from the point where they knew it, hence why they still the great archons did not know who Christ was and the Father himself: "*It was not possible for them to know who the Father of Truth is.*" (2nd Treaty of the Great Set) So the texts speak of the appearance of the Mother by thought and of the Son by the Word, which in Greek is called "Logos". Life would come from this Mother, but always with the consent of the Father. But who is that Mother and why has she never been heard of in the Bible? Certainly it does not seem to be Gaia, Mother Earth, since we are talking about the First Creation, that is, the Spiritual Universe, not the physical world. To address this question, ancient manuscripts have written: «*She is the first Thought, the image of the Spirit. She became the universal womb, because she precedes everything, the common Father, the first Humanity, the Holy Spirit, the triple male, the androgynous with three names, the eternal kingdom among invisible beings, the first to emerge.*» (Secret Book of John 3:8) That is, she was before what would be "the common Father" (the union of the Father and Mother, understood as divinity), she is before the first humanity -the imperishable-, it is before the Holy Spirit, before Christ and before the Eternal Kingdom itself that was formed after it. In other words, the Mother (called Barbeló) would be the personification, name and/or identity of the Thought of the Father.

This force is not understood as a goddess, although mystics and occultists have tried to worship it, as long as they do not give it to the Father or the Son. Thus, Barbeló (Recognition, Previous Thought) requested three gifts from the Creator: Immortality, Eternal Life and Truth, and these five personified concepts (the Creator and the Mother, with their three granted gifts) would have founded the Spiritual Universe. Even the word Barbeló or Barbelón happened, after the Flood, to become a negative term, taking shape in the words: Babeló, Babili, Babalon, Babel or Babylon. Although, she is not given any important role or role, because she simply represents the Thought of the Father, which results in his Word, which is the Son, whom in Greek is called "Christ". Here it is always emphasized that everything was created immediately by the mind and not by the use of technology, magic, copulation or by the independent action of any of them - and evidently under the approval of the Universal Father. After this first Mother, called Barbeló or Barbelón, the Father and the Mother created a thought of light so that now the Son came. This is so, at least as the writings possessed by the first Egyptian Christians refer to it. Then, the recognition or glory of the Creator, which is Barbeló, and the Father united their thought creating a great light, the fruit or "result", which was Christ. From here came the rest of the universe of the imperishable beings of the spiritual world, under the order of the Creator and dependent on Him. Let us understand that this has to do with what the Creator said to Enoch: «If I turn my face, *then all things will be destroyed*." (2nd Enoch 33:5) The Gospel of the Egyptians defines the first creation as manifest from the Incorruptible Father and revealed in three powers: the Father, the Mother and the Son, being manifest from the silence of the hidden Father. This is the reason why the gods of chaos have deceived the world by creating their own idea of three deities: father, mother and son; duplicating, as always, the patterns of the Higher Realms.

The Gospel of the Egyptians or of the Great Invisible Spirit (a treatise attributed to Set, son of Adam), affirms that behind that which was first to show itself came Domedon Doxomedon, from which the later powers arose, so that «the Son a[*rrive*] *fourth, Mother [fifth], [Father] sixth. He existed [...], but without proclaiming. [It is] the one who is insignificant among all [the powers], the glories and the incorru[ptibi]lities.* » (Verse 41). Every time the figure of the Father was identified, an allusion would have been made to this great Kingdom of the 10, speaking of the 8 qualities of this First Creation of the two (Father and Mother), based on the Thought originated from the bosom of the Father (also called Virgin Spirit or Great Invisible Spirit). It may be that despite the parallelism of these events with those exposed in Egypt about the Heliopolitan Ennead, these later ones are early principles that the Egyptian priests documented, but still none of this has to do with anything that has been known on Earth. This all happened *«in silence»*, before all things. Following, more or less, the order of the story, the Son (the beginning of the Light) received the gifts of the Mind and the Will, and these qualities, from above, began to give rise to what is usually called the "Mental Universe": « *Then the Kindred could stand before the powerful and invisible virgin Spirit as God, Christ, produced by himself, whom the Spirit had honored with great acclamations. This Kindred came out through Forethought. The invisible virgin Spirit placed this true, self-produced God above all, and made all authority and inner truth subject to him. Then the Kindred could comprehend the universe that is called by a name greater than all names, for that name will be spoken only to those who are worthy of it.* » (Secret Book of John 14:14-15) We could summarize this as "the word of God made action". If the Father's thought was capable of conceiving such a life, how much more was his mere Word not?

At a certain point, a brother of Jesus, James, wrote, after a debate with the religious Jews in relation to Christ, weeks after his Resurrection: «He was [he] *whom they have not [seen] who created heaven and the land living with it. He was the one [who] is life. He was the light. He was the one who will be. And furthermore, he is going to put an end to what [has] begun and to provide a beginning for what is about to come to an end. He was the Holy Spirit and the Invisible One who did not descend to earth. He was the virgin and what happens to him according to his will."* (2nd Apocalypse of Santiago). It is necessary not to confuse the idea of this Eternal Kingdom with human polytheistic tendencies, such as the idea of "the Queen of Heaven", which is nothing more than a concept of the Canaanites idolizing lesser gods that they considered rulers of the cosmos and its ordering. Although, before the revelations of Christ, this was not openly known, there were only manuscripts in privileged sectors, and furthermore, these texts were not fully understood. The kingdom presented to the Israelites corresponded to the administration of the Material Universe, given to Jesus, but governed by Jehovah. However, in the higher realities, there are existences of pure light that follow a similar pattern but are subject to other canons that do not correspond to physicality. At least that is how it is expressed in the texts found in Nag Hammadi, in the mid-40s, which belonged to the followers of Jesus who settled in Egypt, thus escaping the hands of the polluting Roman Empire. «- *Since the Creator made the whole world, not with his hands but by word, so think of him present and always existing, maker of all things, Unique One, as having created beings by his own will. Because they are truly his Body, intangible, invisible, immeasurable, beyond dimension, incomparable with any other body; for it is not fire, nor water, nor air, nor spirit, but all things from it."* (From Hermes to Tat: the sea, unity. Corpus Hermeticum)

Christ

We see that the concept of the Father, as "God", figures from this concept of three persons, sometimes confused with the trinitarian deification towards the polytheistic gods (an element that we will deal with later), but that it does not include the Holy Spirit. These three presences are also described as matrixes of the "Three Ogdoads", as Set would say: «*From this place proceeded the three powers. The three Ogdoads that [the Father] in (the) Silence together with his Pre-Thought [manifested] from his womb, that is, the Father, the Mother and the Son.*" So, the Great Invisible Spirit, Creator or Universal Father created the figure of the Three Ogdoads that we know as "Father", which places the Silent and Creator above the very concept that many already had about him. If the Father is "*holy, holy, holy*", obviously Jehovah was not the Father but a very low form of his identity, manifest in our world. This leads us to understand that the Father allows himself to be seen in realities inferior to Him, Jehovah having been the first one -although he was really identified in Exodus by angels bearing this name (Chap. 23:20-21)- and the second Christ, who made him known to his people in what has to do with love, mercy, peace, holiness, forgiveness, goodness, meekness, humility, justice and also power. Even with everything, it is promised that God will come down to Earth together with Christ, which is clearly understandable as one more personification of the Father in an exalted way, although there are more above these, all of them in his name and one with Him, but none is Himself – although everything is contained in Him. That is why "God is not jealous", but Jehovah, in the face of those who usurp his place on Earth and other adjacent places. The Son is also mentioned in the Gospel of Thomas Didymus: «*Jesus said: "I am the light that is over all things. I am the universe, [I am everything]: the universe has come from me, [everything came from me] and has come to me. Split a log and there I am; lift up a stone and there you will find me."* (Thomas 1:75)

If indeed Christ came from this Higher Kingdom, that explains why he said: "*Fair Father, the world has not known you, but I have known you, and they have known that you sent me. And I have made your name known to them, and I will make it known again, so that the love with which you have loved me may be in them, and I in them.*" (John 17:25-26) When saying "name" identifies the Hebrew voice "shem", which refers to a "renown", "identity", "reputation" and meaning of what the person himself is and what represents, not necessarily an identifier or title. That is why Jesus revealed his "shem", that is, what God is, represents, means, etc., not a specific name as an appellation, as it is usually understood in religious monotheism. The Egyptian text of Seth, which speaks of the Father, adds that " *then [the Infant] thrice masculine [of the great] Christ who had been anointed by the [great invisible] Spirit, the one [whose] power was [called] Ainon, gave praise to the great invisible Spirit* " (vers. 44), and asked him for virtues, which he received, as well as 12 Kingdoms over which he would be sovereign, starting from 4 aeons: "*He established thrones in glory, innumerable myriads in the 4 aeons that surround him, innumerable myriads, powers and glories and incorruptibility. And it arose in this manner.*" (Vers. 54-55) Accommodating these words of Set to his second manuscript, Christ seemed to have intervened in the affairs of the universe, in different parts and times, where the physical body was nothing more than a bridge —something that we will clarify later. We can add another passage in which Jesus talks about himself and his role: «*Matthew asked him: "Lord and Saviour, how did humanity appear to you?" And the perfect Savior answered him: "You must know that the one who manifested himself before the endless universe is the one who has believed by himself, who has built himself, the Father full of radiant and ineffable light. In the beginning, he decided that his image would become a great power. And the principle of this light immediately manifested as an immortal and androgynous man, and, thanks to this immortal Man, they will be able to ensure their salvation and be awakened from oblivion by the intercessor who was sent to them and who is with you until the end of life. poverty of bandits.*" (Sheneset Gnostic 1:10).

It is appreciated in the Scriptures that Christ is the Firstborn of the Father and is his Only Begotten, but few stop to think what that means. It is also designated as light, but what does that mean? Jesus, exposing this to his disciple John, told him: «*The Father penetrated Barbelo with a look, with pure, brilliant light, which surrounds the invisible Spirit. Barbelo conceived, and the Father produced a ray of light that resembled the blessed light but was not as bright. This ray of light was the single Offspring of the common Father that had come forth, and the single offspring and single Offspring of the father, the pure light.*" (Secret Book of John 4:1-3) If this light was the first to be produced, and it is identified with Christ, then we can appreciate that there are biblical passages consonant with the theme, not only in Genesis 1:3, but in the New Testament –since the light of Genesis 1:3-4 came because of darkness, that is, it came after the light of the Son-, as for example: «*That true light, which enlightens every man, came into this world.*" (John 1:9) But we understand that the definition of light can also be allegorical, implying that it is the clarification on all things, the pure truth and complete clarity. In any case, both things can be closely related: «*And this is the condemnation: that light has come into the world, and men loved darkness rather than light, because their works were evil. For everyone who does evil hates the light and does not come to the light, so that his deeds may not be rebuked. But he who practices the truth comes to the light, so that it may be manifested that his works are done in God.*" (John 3:19-21) Because of this we see that John Zebedee identified John the Baptist as a preacher of Light (John 1:7-8), and Jesus referred to him as a giver of momentary light (John 5:35). We therefore have that Jesus represents the Light that comes into the world, possibly in association with the identity that he has as the manifest light since the origins of everything, before the visible existed. This is how he presented himself to the Jews: «*Again Jesus spoke to them, saying: I am the light of the world; whoever follows me will not walk in darkness, but will have the light of life.*» (John 8:12. Similar to John

9:5). Whereupon the description is, in this case, about ignorance, but also shows it as light itself.

However, the clearest appreciation of his relationship with the true Light and knowledge of all things, at the same time, we also see in the Gospel of John, when he said: «The light is still with you for a little while; *walk while you have light, so that darkness does not surprise you; because he who walks in darkness does not know where he is going. As long as you have the light, believe in the light, so that you may be children of light. These things Jesus spoke, and he went and hid himself from them."* (John 12:35-36) Jesus presents himself as light, this being a punctual symbol of how he was before coming into the world and how he brings the truth in the midst of ignorance. This is reflected in contravention of some religious denominations, which suggest that he was Jehovah or Jehovah's angel, depending on the doctrine, making it clear that he was not on Earth before he was born but in the bosom of the Father (John 7:28, 1 :1-3 and 1 John 1:2), being expressed especially as the personification and representation of "Life", that is, Eternal Life. In addition to all this, he is credited with being the engine of the origin of man, hence the nickname "Son of man" (Ben ha-Adam), not for being a "person" (Ben Adam = son of Adam) or " descendant of Adam" (Bnei Adam), but for being the very representative of the human race, not only terrestrial, but the imperishable one, which precedes it. The cited texts are worth studying carefully, because, although they are complex, they reveal great things of edification, including explanations on this matter of "Christ" and the "Son of man". Reading we could understand why Jesus said: « *My kingdom is not of this world; if my kingdom were of this world, my servants would fight so that I would not be handed over to the Jews; But my kingdom is not from here.* » (John 18:36). As the writings, especially Gnostic (from 1st century Egyptian Christianity) seem to explain, Christ is sovereign already from those beginnings, and was given great power and assistant leaders under him. Even, with the collaboration of the rest of the great Assembly, it gave rise to the first immovable race called Adam, and the birth of the great Seth, but not the terrestrial figures that bear these names, but the celestial ones, for which he would receive the name of Son of Man, that is, image of the original design that represents what we know as "human being".

Yet he was, and is, one of the many facets of the Father's light, both manifest and unmanifest, not merely the Messiah of Israel or the gatherer of the lost tribes (Isaiah 49:5-7). Enoch wrote about the Eternal Son: « *At that moment that Son of Man was named in the presence of the Lord of Spirits and his name before the Head of Days. Already before the sun and the signs were created, before the stars of Heaven were made, his name was pronounced before the Lord of spirits.*" (1 Enoch 48:2-3) And he added: " *For this reason he has been Chosen and reserved before Him, from before the creation of the world and forever.*" (1st Enoch 48:6) Enoch confirms what the Son would be when he came into the world, saying that « *He will be a staff for the righteous on which they can lean and not fall; It will be light for the nations and hope for those who suffer.*" (1 Enoch 48:4) He also prophesied about the Ministry of Repentance, in this way: «*In the day of affliction, when misfortune is heaped upon sinners, the righteous will triumph by the name of the Lord of spirits and will make Others testify that they can repent and renounce the work of their hands.* » (1 Enoch 50:2). Along the same lines, Enoch also mentions the knowledge, in advance, of those who would be chosen to dwell with the Exalted: « *During the same days the Chosen One will rise up and select the righteous and the saints from among them, because the day is near. day when they will be saved.*" (1 Enoch 51:2). Likewise, thousands of years before he was born in Bethlehem, he warned about when Jesus was glorified and received all power: «*The Chosen One will sit on his throne in those days and from his mouth will flow all the mysteries of wisdom and advice, because the Lord of Spirits has granted it and has glorified it.*» (1 Enoch 51:3)

The Undying Realms

There is much more to say about the Son, Christ, and about the Father, but we must also deal with other points, such as the forms of life that appeared in the Kingdom of the Father (the other details can be found nuanced in "Attaining Deity" and "A Forgotten Message"). John wrote that Jesus told him: "*The Perfect One contemplates himself in the light that surrounds him. This is the spring of the water of life that produces all the worlds of all kinds.*" (Secret Book of John 3:1) Jesus said that he was not of " *this world* ", which is why he came from another place, or from another world. John was not the only one to refer to other worlds, which are created and maintained by the Father: « *The worlds are infinite and man has to live in all those that exist today; but the creation continues and does not end. All the worlds communicate with each other in love and justice, and my God is magnified in it.*" (Excerpt from the Secret Testament of Abraham 1:2-3, dated 4,000 years ago) Would these worlds already exist before the establishment of the universe? Set wrote that " *before the foundation of the world, when all the multitude of the Assembly gathered in the places of the Ogdoad*" (2nd Treaty of the Great Set), the next destiny of Creation was determined. Certainly, before these worlds appeared as such, there existed the powers arising from the Universal Father, who received special qualities and assistant creatures. After this came the Kingdoms created as housing to dwell there. Emergent qualities seemed to always be personified. The first to request company for her was Barbeló (the Previous Thought), who asked for Previous Knowledge, Immortality, Eternal Life and Truth. This first step was the one that gave rise to the «*kingdom of 5 of the Father. It is: the first humanity, the image of the invisible Spirit, that is, the Previous*

Thought, Barbelo, Thought, together with Previous Knowledge, Immortality, Eternal Life and Truth. This is the androgynous kingdom of the 5, this is the kingdom of the 10, this is the Father." (Secret Book of John 3:17-18) Here it is explained that when referring to the Father one speaks of this Kingdom of 10, which has in it the Kingdom of Barbeló, which is of the 5, which is identified as the first humanity to arise, that is, the first created.

About the 4 Eternal Kingdoms, and the fourth, destined to the Son, another text says: « *Then the Savior created one [only] after all of them, and the minds of these read, the blessed and different because of their choice; and others, many of those who are without a king and all those who precede them, so there are 4 races. There are 3 that are assigned to the kings of the Ogdoad, but the 4th race belongs to the king and is perfect, because it is over all.* » (Apocryphon of Coptic Creation 124:33 to 125:7) Those 4 were guarded by 3, which in turn govern their own system, with which there are 12 Base Imperishable Kingdoms. However, the attribution of kingship should begin by offering rule to Christ. This is how the Kingdom of the Son came, not without first being anointed by the Holy Spirit, when he manifested himself for the first time as light: « *The Father anointed her with goodness until she was perfectly and completely good, for the Father anointed her with the goodness of the invisible Spirit. The Kindred was in the presence of the Father during the anointing."* (Secret Book of John 4:5) What Christ asked for was Mind, and so it was. Then came the Will, but all according to the power and design of the Word of God that was in harmony with the Father and through which all things arose. So Christ, ergo, received power and dominion over all: "*Then the Kindred could stand before the mighty and invisible virgin Spirit as God, Christ, producing of Himself, whom the Spirit had honored with great acclaim. This Kindred came out through Forethought. The invisible virgin Spirit placed this true, self-produced God above all, and made all authority and inner truth subject to him. Then the Kindred could comprehend the universe that is called by a name greater than all names, for that name will be spoken only to those who are worthy of it.* » (Secret Book of John 4:14-16) So then, like Barbelo, he also received a kingdom and the corresponding helpers: «*Now from light, Christ, and from Immortality, by the grace of the Spirit, came the four stars that are derived from the God produced by himself. He looked around him and made the stars stand before him.*" (Chap. 5:1) As we can see, no one created by themselves, no one did things alone, all worked together to give rise to forms of life, powers, creations, and other manifestations. This reminds us of the luminous creatures that Enoch observed on his first trip: " *Then they took me to a place whose inhabitants are like burning fire, but when they wish they appear as humans.*" (1 Enoch 17:1)

You can imagine that this whole realm is nothing more than planets and systems in outer space, but even so the Father is not subject to anything that is observed with telescopes. The Father, defined as Virgin Spirit or Universal Father, is unique. However, like a body, it has many parts, but they are all He and part of Him, since man is the one who always sees Duality and Separation, distancing himself from Oneness. The Virgin Spirit, the Universal Father, just raise a Thought, it was already personified (this is the degree of its unlimited power), and that Thought, called Barbeló, together with the Universal Father, created the Son, who is not another more than the personification of the Word that came out of the Father. All this that began to be revealed received the personified name of Domedón Doxomedon. When they exist they receive qualities and companions, with which they organize their Assemblies of various numbers of components, being worlds, kingdoms and unity with the Father, all at once. All this team in general is defined as Holy Spirit, or simplified as Spirit, in consonance with its strength, motivation, movement, activity and unanimity in the holiness of the Father. At the same time, those who receive the concrete authority, receive the ability to be understood as "the Father", even if it was not exactly the Virgin Spirit, since they express what the Father is and are one in themselves, since in them there is no separation. no difference of opinions or critcria, no divisions or interests or different essences. Thus, they show the true union, where all, by common agreement, place things, and are understood as "The Father" and as the real deity, whom we understand, from Christ, as "God" -whether we Let us refer to the Virgin Spirit to Christ, to Barbeló or to any other, since, I repeat, there are no differences between them, nor jealousy, nor envy or criteria of individuality, since these "criteria" have precisely caused human isolation.

In this way, beings and powers came together to create, but not before the stars came to Christ, which represent great celestial eminences or luminaries that in the world we call "archangels", and who could equally symbolize stellar systems. This is how *"the power of the great light, the manifestation, came from above. She generated the 4 luminaries: Armozel, Oroiael, Daveité and Elelet and the great incorruptible Set, the son of the incorruptible man, Adamas."* (Gospel of the Egyptians. Vers. 52) Then thus " *the perfect Hebdomad that exists in hidden mysteries was completed."* Each one of these luminaries was placed on one of the 4 Great Eternal Kingdoms, but helped by a collaborator. The personified beings and powers also appear to be Realms themselves or representations of Realms. So even though there are 4 archangels on this, there are actually 12 imperishable realms. The first of the leading angels is Armozel, whose accompanying qualities were Grace, Truth, and Form. The second is Oroiel, accompanied by Afterthought, Perception and Memory. Then there is the third, Daveitai, with the realms of Understanding, Love and Idea. And finally there is the kingdom of Elelet, where there are also Perfection, Peace and Wisdom. Thus said the Lord: « *These are the 4 stars that are before the self-produced god, and the 12 eternal kingdoms that are before the great Offspring, Christ, produced by himself by the will and grace of the Spirit invisible. The 12 kingdoms belong to the self-produced Kindred, and therefore everything was established by the will of the Holy Spirit through the self-produced.*" (Secret Book of John 5:7-8) This leads us to understand why on Earth there are also 12 representatives of each one of the tribes of Israel: what is below is an image of what is above. And, likewise, we see that there is a Christ above and another below, a Father above and another below, and even a Holy Spirit above and another below: reflect them one from the other, but in consonance and without perception of usurpation, but of unity and affinity Complete: ONE. Therein lies the depth of Jesus' words when he clarified: "*My Father and I are one.*" (John 10:30)

Therefore, it was proposed that there would come, emerging from immeasurable power, attendant angels as companions, to which the Father agreed: « *The first the great Gamaliel, (of) the first great luminary, Armozel. And the great Gabriel, (of) the second great luminary, Oroiel. And the great Samio, from the great luminary, Daveite. And the great Abrasax, of the [great luminary], Elelet. And [the con]sorts of these proceeded by the will of the good will of the Father, the Memory of the great, the first, Gamaliel; the Love of the great, the second, Gabriel; the Peace of the third, the great Samblo; the Eternal Life of the great, the fourth, Abrasax. Thus the five Ogdoads were completed, forty in all, like a power that cannot be interpreted.*" (Gospel of the Egyptians 1:52-53). So Gamaliel helps Armozel, Gabriel helps Oroiel, Samio (Samblo) helps Daveitai (Daveite), and Abrasax helps Elelet. Jesus also spoke to the group of followers about this, saying that "*the Autogenous Father began by creating 12 aeons for his succession and 12 angels*" (Sheneset Gnostic 1:15) "*unlimited abodes of the immortals*", he replied that « *the first aeon is that of the Son of Man, the so-called Savior, the one who has been revealed, the one who has been called the First Genitor. The second aeon is that of Man, who has been called Adam, the Eye of Light. The one that surrounds them both is that aeon over which nothing reigns, that of the divine and immense eternity, the autogenous aeon of the aeons that are in it, that of the immortals of which I have spoken before, the one that is above the seventh revealed by Wisdom, which is the first aeon."* (Sheneset Gnostic 1:16) We must bear in mind from now on that the translators define Wisdom by its Greek meaning "Sophia". She will be a crucial aspect in our work, because it depends on her that we are here and that we are dealing with this issue. Precisely the aspect of the "Spirit" of the Lord, God, Jehovah or Elohim concerns a "collective". It is not appropriate to consider that the Spirit is not linked to being or acts separately from it, but rather that the definition (also equivalent to wind, force, energy, equipment, motivation and essence) welcomes a "common" understanding and work. That is why the messianic prophet wrote about what God revealed to him: « *Come close to me, listen to this: from the beginning I did not speak in secret; since that was done, there I was; and now the Lord GOD sent me, and his Spirit."* (Isaiah 48:16) In other words, the biblical concept of "Spirit" is also the abbreviation to refer to the heavenly, excellent and in harmony and consonance with the father, while in turn expressing what the presence of God is.

According to Abraham's aforementioned, the worlds already intercommunicate, apparently, this was already the case before Chaos began: «But he, the *immortal Man, has revealed aeons, powers and kingdoms granting, to all whom he revealed, the power to fulfill wishes until the last times before the chaos. For they have understood each other."* (Sheneset Gnostic 1:17) He further adds that « *all greatness has revealed, even by the spirit, a multitude of glorious and innumerable lights, those chosen from the origin, that is, the first aeon, and the second and the third. The first was called Unity and Rest; and each of them has its name. For they have been called the Assembly of the three aeons among the innumerable multitude that at once appeared. And a multitude has revealed to him. But since these multitudes have come together making a unit, they were called the Assembly of the eighth. That it has been revealed androgynous and has been partly named as male and partly as female. The male has been called Assembly, while the female has been called Life, so that it would appear that the life of all the aeons has issued from one woman. And each of those names from the origin would be received.* »

Then he points out that for his pleasure these first powers were called gods. So, given this, it is difficult not to wonder: Does this have something to do with the Ennead mentioned in hieroglyphics and Egyptian texts? The said Ennead mentions 8 primordial gods (4 couples) and a demiurge. As you will have read, the name of Ogdoada, to refer to a group or assembly, is clearly Greek, since it defines a group of "eight", and thus there are several guilds that make up meetings representing the Most High.

    « *There was a wise man, a great artificer, and the Lord conceived love for him and received it, for he must behold [the] abodes of the Most High and be a visual witness of the Wise and Great and [of the] inconceivable and unchanging realm of Almighty God, and of the most wondrous and glorious and bright and [the] many-eyed stations of the servants of the Lord, and of the inaccessible throne of the Lord, and of the degrees and manifestations of the incorporeal hosts, and of the ineffable ministration of the multitude of elements, and of the various apparitions and of the inexpressible songs of the hosts of the Kerubim, and of unlimited light.* » (Introduction to the Apocrypha of Enoch, Second Principal of Enoch or Slovak Book of Enoch) All this primeval conglomerate, apparently called "Pleroma", once it completed the first phase of creation, advanced towards the bottom of it, passing through Youel and Esefec, other beings and the so-called "Five Seals". So more life took place, because the light of Christ and the immortality of Barbeló, by the grace of the Father, created the first 4 angels, lights or stars, to initiate the leadership and administration of the first 4 imperishable kingdoms, as we have said. These 4, together with the ruling 3 of each one, are in total the 12 Kingdoms that are in front of Christ and belong to him. He, Christ, governs and governs them with his power, called Ainon, and they are called as a whole the "Hebdomad". These initial worlds would be assigned to humans, the race of Adam in different orders.

Set related it in this way: « *They gave praise to the great Word, the divine Self-begotten and the incorruptible man Adamas, they asked for a power and strength forever, for the Self-begotten, for the fullness of the four aeons, so that through them manifest [...] the glory and power of the invisible Father of the holy man of the great light that will come to the world that is the appearance of night. The incorruptible man Adamas asked them for a son from him, so that he would be the father of the unbreakable and incorruptible race so that through it the Silence and the Voice would manifest and through it the aeon that is mortal would arise so that would* dissolve *[...] And thus the perfect Hebdomad that exists in hidden mysteries was completed. Once he re[ceived] the g[lory] he was eleven Ogdoads. And the Father answered affirmatively. The entire Pleroma of the luminaries agreed. His [con]sorts proceeded for the completion of the Ogdoad of the divine Self-begotten: Grace, of the first luminary, Armozel; the Sensitivity of the second luminary, Oroiel; the Intelligence of the third light, Daveite; the Prudence of the fourth luminary, Elelet. This is the first Ogdoad of the divine Self-begotten.*» (Secret Book of Set, Gospel of the Egyptians or Book of the Great Invisible Spirit) Some of these ancient texts have concepts in a different order but keep the same similarity around the first creations. So, were there two Adams? At times it seems to have been the same, while at other times a representative of that first, because it was written: " *He came down from on high to remove the deficiency.*" (Secret Book of Set) What deficiency? Well, what is being put forward is that all these imperishable beings and the Ogdoad are above the gods of chaos: " *These, in fact, are in their father's holy place, where they enter and go to rest in a rest from the eternal glory and indescribable and endless joy. And they are the kings of mortals as a race of immortals; [for] the gods of chaos and their faculties must be damned.*" (Coptic Creation Apocryphon 125:7-14) We will go into this point later.

On the other hand, everything mentioned did not occur on Earth, nor on this dimensional plane: «*... and the ethereal earth, the recipient of God, the place where the holy men of the great light receive the image, the men of the Father, of the living silent Silence, the Father and the total Pleroma ...*» (Secret Book of Set) We can later read about the decision to create the worlds and empower star-leaders to administer them. And this first universe was made in such a way, as Enoch related –which we will see later-, of what was not physical and still survives. This is how Set defined it in his first treatise: «*Then everyone stirred and trembling dominated the incorruptible. Then the three times male Infant proceeded from above downwards in the unbegotten and the self-begotten and those that were begotten in what is begotten. Greatness proceeded, the total greatness of the great Christ. He established thrones in glory, innumerable myriads in the 4 aeons that surround him, innumerable myriads, powers and glories and incorruptibility. And it arose in this manner.*" So Set sowed his seed with the help of Plesitea, the mother of angels, in the 3rd [and] 4th Kingdom. So, Adam, were you aware of everything that had happened? Yes. Were you aware that you had come to this world from that previous universe? Did he know why he had come to the material world? He himself reiterated to his Son Set: « *Then God, the sovereign of the eras and powers, divided in anger. Then it became two eras. And the glory in our hearts left us, me and your mother*

Eva, along with the first knowledge that we breathed into us. And [the glory] fled from us, [because] it entered into a great [kingdom] [from] which it had come out, not from this Aeon from which we had come out, I and your mother Eve. However, [knowledge], entered the seed of a great time. For this reason, I myself have called you by the name of that man who is the seed of the great generation or of which (it is about). After those days, the eternal knowledge of God's truth withdrew from me and from your mother Eva. Since then, we learn about dead things, just like men. Then we recognized the God who created us. But we [were] not outside their powers. And therefore we serve him in fear and slavery. And after these things, we became darkness in our hearts. Now I lay down in the thought of my heart." (Apocalypse of Adam 1:3) In other words, for that reason Adam called Set the same as the figure of the mental universe. But then, would they be the same coming from another world-dimension or would they be a figure of those who were "above"?

Abraham also commented on this, stating: «*My one God, Universal Creator, has no beginning: he is eternal. Men are his children and he his inheritance. The worlds are infinite and man has to live in all those that exist today; but the creation continues and does not end. All the worlds communicate with each other in love and justice, And my God is magnified in it. All the sons of my God, whom you call angels, were men: because I spoke with Noah, who looked like an angel; because I spoke with Adam and he looked like an angel; because I spoke with Eva and I saw her give birth to a savior and she is a son of my God, who already lived in another world. I am from the race of Adam and my children are from the race of Adam, <u>who have to save the first race that populated the earth</u>; Because Adam and his family came with light and wisdom from my God."* (Secret Testament of Abraham 1:1-5) Are the worlds infinite? Abraham seemed to know about life in the cosmos, how or why did Abraham suppose that men must live in all the worlds that exist, and that they never end? Did Abraham take it for granted that people on other worlds communicated with each other and lived by the same principles of peace and harmony? Abraham writes verbatim that Seth " *already lived in another world* " and that the destiny for which Adam and his descendants, passing through Abraham, had the purpose of having come to " *save the first race that inhabited the Earth* ". is that? Adam was not the first, and Abraham knew that perfectly, but it would be necessary to find out who they came to rescue.

The First Man

Now, when we talk about the appearance of the first man, we are not referring to the human Adam of Mesopotamia, the one who appeared to the east of the Garden of Eden a few thousand years ago, but to the one of said Universe or World from where everything comes. Thus, according to the first book of Seth and the Secret Book of John, the Truth, which is the first revelation and the first man arose, was born from Mirotoe (Mother of holy incorruptibility), and this offspring received the name of Adamas, Geradama or Pigeradamas. Adamas was named for the kingdom of Armozel, the First Kingdom, where Adamas received the power of Unconquerable Thought. In turn, Adamas named a son, Set, for the Second Kingdom, the kingdom of Oroiel, and for Set's family the Third Kingdom, that of Daveitai, would be destined. The latter would be the destination of the souls of the saints. But a Kingdom was missing to accommodate: « *The souls of those who were ignorant of the divine Fullness were destined for the fourth eternal kingdom. They did not immediately repent, but continued to ignore it for a while and then repented later. They are with the fourth star Eleleth, and they are creatures that glorify the invisible Spirit.*" (Secret Book of John 5:16-17) The sacred texts of the Egyptian Library of Nag Hammadi say that the Ethereal Land brought the Four Kingdoms from where the first living beings began to swarm. This Ethereal

Earth would be the "Receiver of God", the place where the holy men of the Great Light receive the "image". I speak of the spiritual, or mental, universe of the Imperishable, those who were before humanity that came to our material orb. This is where Adamas came from, who, joining the Word, created a man-word or word-man union, that is, the word of God made action through the figure "man". All this would be taking place, as Baruch said, before Time. In other words, the "non-time" can be the stay before the guidelines of the periods were set or, simply, refer to the existence outside of our reality, which is subject to the established law of space-time.

Upon resurrecting, Jesus told his disciples disconcerting things about this first Adam (Hebrew definition, Adam, which translates as "man"): «*His disciples said to him: "Lord, he who is called Man has revealed to us on this matter, so many that we too now know exactly his glory." The perfect Savior answered: "Whoever has ears to hear, let him hear. The Father, the First Parent, is called Adam, Eye of Light, because he has come out of the brilliant light with his holy angels, ineffable and shadowless. Who rejoice in a perpetual jubilation of the reflection that they have received from their Father: there resides the kingdom of all the Son of Man, the one who is called the Son of God. There reigns an ineffable and shadowless joy, a perpetual jubilation, for all rejoice there in his imperishable glory, hitherto unheard of and unrevealed to the aeons to come after them and their worlds. From the Autogenous I come and from the primordial and endless Light, that is why I can reveal all things to you.*» (Sheneset Gnostic 1:13) Jesus seems to make it clear once again that he came to represent all that is Above, including the celestial Adam, Father of the human race. However, the fruit of the celestial Adam also had children and placed them in the corresponding abode, "*he asked for his seed*" (Egyptian Gospel 1:56), "*then came from this place the great power of the great Plesitea luminary, the mother of angels, the mother of lights, the glorious mother*", she "*came through the great Seth*" and thus, from her, he "*took his seed from the 4-breasted virgin and placed her with him in the 4th aeon [and] in the 3rd great luminary Daveité.*" After these events, everything seems to have begun to change radically. About them Enoch could have made a description when he said that, when he was taken to dwell in the Heights, he saw the first humans: "*There I saw the first parents and the just who from the beginning dwell in that place.*" (1 Enoch 70:4)

If this was "before time" it means that it was not taking place in this universe (definition meaning "all verses" or "various versions", although some believe it means "single verse" or "single version"), since that it is governed by the strict law of "space-time" and to it all forms of life that exist here are subject. Humanity then owes its origin to higher forms of a reality beyond the one we know. As the American archaeologist and researcher, Michael A. Cremo, said in a television interview decades ago, after many years of detailed studies and work: " *human beings are not native to this planet,*" adding that "*we came from another dimension... another universe.*" It is not uncommon for other speakers and researchers who have been on many of these topics to suggest the same idea, even in the area of ufology. The terrestrial Adam and his son, the terrestrial Set, would have been aware of their origin, but after centuries, the words from father to son would gradually dissipate and be completely lost. It is now possible to inquire into certain existential questions, such as the fact that man does not remember where he comes from. The question of whether, besides Adam and Set, there were other people from said universe who descended to this plane also comes to mind. Likewise, the entire conceptual idea of the past, the present and the future would be reconsidered in another way if it were admitted that humanity has deliberately come to this planet from another universe, since, if so, what objective would there be? Since we accept that man experiences hardship, pain, disease and death, and it is illogical that he was a masochist. In the same way, he would not have come deceived, not knowing what would be in store for him here. It is necessary to dust each and every one of these elements and answer such basic questions.

«- And what need was there, oh Trismegistus, to put man in the material world and not in that part, where God dwells, and who lives in supreme happiness? - How well you question, oh Asclepius! and we pray to God to grant us the power to explain this subject to you. As all things depend on his Will, both it and the things that refer to the entire Sublimity, are the matters whose explanation we seek. Listen then, Asclepius. The Lord and Maker of the Universe, who we rightly call God, who made a second god that could be seen and touched, - second god that I called "sentient" not because he feels (of which, whether he feels or not, we will say in other place) but because it falls under the sense of those who contemplate it - when, then, God, from himself the first, had produced this second and had seen it beautiful, since it fully contains the goodness of all things, he loved it as part of his divinity. And then, as Almighty and Good, he wanted to make another one that could contemplate the one he had taken out of himself, and immediately makes Man, an imitator of his Name and his Diligence. The sole Will of God is the highest Perfection, in such a way that in the same and unique moment of time his will and his accomplishment coexist. As he made man OUSIODES and understood that he could not take care of all things if he did not place it within a material texture, he wove a corporeal residence for him and ordered that all men be composed of both natures, confusing and mixing them as much as necessary. Then man was made up of soul and body, that is to say, of the eternal and mortal nature, in such a way that, conformed like this as a living being, he could give satisfaction to his two origins: to look at and adore celestial things, and to cultivate and govern the heavenly things. earthy.» (From

Hermes Trismegistus addressed to Asclepius, verses 7-8. Corpus Hermeticum) The word Ousiodes is a Greek definition that would mean, in Egyptian mentality, "figure of divine likeness."

The Preexistence

If Adam and Set existed "before they existed", that would give rise to the theory of Pre-existence - something very different from the theory of Reincarnation - of which Jesus is said to have spoken, but whose parts would have been eliminated from the Bible in Roman Catholic councils. For example, this fragment of the Gospel of Thomas Didymus points out something that pushes this belief: «*Jesus said: "That __the flesh has come to be thanks to the spirit__ is a prodigy; but the fact that the spirit (has come to be) thanks to the body, is a prodigy [of wonders]. And I marvel at how this great wealth has come to reside in this poverty."*» (Thomas 1:29) It is evident that the human body is a bond, a vehicle or mass that does not correspond to the essence of the imperishable universe although visually it is similar. We can again read about this statement a link from Abraham that says: «*... I spoke with Adam and he looked like an angel; because I spoke with Eva and I saw her give birth to a savior and she is a son of my God, who already lived in another world.*" (Secret Testament of Abraham 1:4) Abraham says that "they were angels" and that Set " *already lived in another world* " before coming to ours, as explained above, but, as all these writings point out, there is a "forgetfulness" which has been placed in the race of men so that they do not know or remember where they come from, which was not established by the Most High, but by a common enemy . Jesus himself mentioned many things in association with pre-existence in the Realms of Light. In a specific case, speaking to his disciples, «*Jesus said: "If they ask you: Where did you come from? Tell them: __We come from light, from the place where light originated by itself; (there) she was settled and manifested in her image__. If they ask you: Who are you?, say: We* *are his children and we are the chosen ones of the Living Father. If you are asked: What is the sign of your Father that you carry in yourselves?, tell them: It is movement and rest at the same time.»* (Thomas 1:50) The Lord seemed to suggest over and over again that characters like Adam, Eve, Seth, Noah, Jesus, the 12 apostles and many others deliberately came from another universe with the purpose of rescuing the rest of mankind rooted in this orb. It is not specified if the entire line coming from Adam is, in fact, coming from the Eternal Realms, but it is made clear of those who have come with specific functions.

Ergo, the theory of pre-existence -I repeat: let's not confuse it with reincarnation- supported by Gnostic texts, assumes that before the creation of the Material World all the spirit-souls were with the Creator, in its bosom. This philosophy says that the Universal Father created the Spiritual Universe and later the Mental Universe where the Children of the Universe, or imperishable beings, lived. Some ufological theories treat this matter and see it in a similar way, although they add that humans and extraterrestrials would have existed in other universes before at will, and climbed the steps of learning in other lives before the existence of our material universe that we know, the plane that touched them or the planet. According to all this postulate, the imperishable of the "Adam" genre wanted to live in a new Material Universe subject to "certain laws", incarnated in physical bodies where they would ascend spiritually and learn from scratch, with which the objective of the life of beings of conscience -us- is to LEARN. The Greek Plato also spoke of this idea, of being before being, when he put forward the theory of Reminiscence, which presupposes the pre-existence of the soul, since if "knowing is remembering", somewhere the soul had to contemplate what is now remember. If it existed before being in this body, it will exist after leaving it, since there will be a circular journey from life to death and from death to life, as there is between all opposites (big little) –according to Plato. If death is decomposition, only the compound, that is, the body, is corrupted and dies; the soul, being simple as ideas, is incorruptible and immortal, since nothing in the universe is destroyed but only changes its form.

If we look in the direction of what Christianity calls "the Word of God", the biblical story, this was also discussed, however, the Roman Emperor Justinian introduced innumerable changes to Christian doctrine (by then better defined as "Catholicism"). He called the Second Synod of Constantinople in 553 AD « *The Synod was neither presented nor apparently sanctioned by the Pope in Rome. In fact, at this time, many of the changes to the position of the Christian creed in the Eastern Roman Empire had not reached the papacy, although they did eventually.* » (William Bramley, Gods of Eden) The Second Synod issued a decree excluding the idea of "past lives" or "pre-existence", although apparently said postulate was important to Jesus, according to the Gnostics. The Synod decreed: "*If anyone affirms (supports) the fabulous pre-existence of the soul and yields to the monstrous doctrine that derives from it, make him anathema (excommunication).*" Jesus had taught that each human being is responsible for his destiny, "*but in 553 the suppression of those teachings was voted, with the purpose of consolidating the control of the [Catholic] Church, which wanted to be the sole authority on the matter concerned.*" to the fate of man." (William Bramley, Gods of Eden. Emphasis added) The then Church-State concordat, frightened to see that a doctrine that makes individuals responsible for their own spiritual salvation could confront their authority, coined such a prohibition. « *Those who claimed to represent Christ did not want the people to know that they did not need creeds or rituals to reach the Kingdom of the Father. Those leaders were not interested in letting people know about karma, the law of "cause and effect." [Some have hypothesized that even this suggested that] wrong conduct could bring ever greater responsibilities to the incarnated spirit. Consequently, the Catholic leadership cut the paragraphs where it explained that, to reach the Father, only altruistic behavior was enough. No one then needed a religious monopoly to drive them like a herd."* (William Bramley, Gods of Eden. Emphasis added)

I am not stressing Bramley's words with the intention of suggesting that "living in Christ" is not important, since it is the most important thing, I am saying rather that it is not the Church (thus understanding the Temporal Power that the Vatican has given itself) the bridge to reach God nor is idolatry. Likewise, the fact of the Resurrection of the dead has to do with the Father and the Son, not with the Church (whether in Rome or in any of the aspects of popular Christianity). Therefore, *"that doctrine was reinforced with promises of eternal glory to those who submitted to [Catholicism] and eternal fire in hell to those who rebelled against the established concept. The feeling of guilt was manipulated and the conscience was tamed, leading to delegate personal power to the curia (church) itself, which had united with the Emperor Justinian and, among themselves, woven the greatest conspiracy against the spiritual path »* (William Bramley, Gods of Eden. Emphasis added), assuming the effective Resurrection of the flesh -although not even they really visualize it as it is-, but adding an ascension to heaven after the death of the just, men being entirely dependent on the monopoly of Roman Catholicism, not man's own way of following the words of Jesus. Although, the true Resurrection is the Resurrection to the Light, and this is where most beliefs err by overlooking or not understanding the universal end. Some ufologists maintain, in reference to words apparently received by extraterrestrials, that *« They affirm that the Supreme Maker has plans for this planet. The Earth is a chosen planet, and the mistake of many spiritual centers is to think that this planet is a planet of punishment, a planet of error. They believe that we come to this planet to learn."* (Jorge Olguin, ELRON group). This can lead to lengthy debates, based on questions like: if this is so, then why do we have to suffer? Is it not possible to learn without the need for so many sufferings? And this is true, but there is "fate" and "evil" that got involved in the path of human life; something that we will mention in detail later.

If man comes from another place, do we all come from there? The Gnostic texts affirm that Jesus spoke of those who "believed", as if he were dealing with those humans who, in the saga of "The Matrix" films, awaken from virtual deception. It can be assumed that those who reach the end do so because they have really been one of the imperishable incarnated in this world, although they did not know it for sure. In other words, those who recognize their identity would be saved, which is linked to what it implies: a healthy life, in justice and with good deeds. True or not, this does not change the facts, because whoever reaches Salvation will have achieved it, and whoever does not, well no. Whether they find out about this or not, because the point is that the message of Salvation has reached the world in such a way that each one decides whether to accept it or reject it. Now, the idea of being on this plane has been accompanied by approaches in relation to the "substance" or "essence" of this same cosmos. In other words, it has been suggested that the universe we know, the observable, is a holographic projection to where we have stopped to live. Added to this is the idea that the reality we experience is the embodiment of our thought seen in retrospect or felt in this way and generating life. If so, the question does not stop being reflected: Why did we fall here? Who created this universe and for what purpose? The conception is repeated in some other areas of research, where they maintain that the cosmos is an unreality that only existed in our own mind, which would have created it from the bosom of the Father, being us part of it. Could we accept it? This is how Buddhists and Hindus see it, the unreal universe of Maya.

There are ideas that are very difficult for many to consider, like this one, but the writings are there, and they reiterate what Abraham would have said at the time and what even modern science has begun to discover. I remember what was referred to in other post-biblical texts: «*[Adam said:] Listen to my words, my son Seth. When God had created me from the earth, together with Eve, your mother, I went with her to a place near a glory that I had seen in the <u>previous world-century from which we had come</u> [...] And a glory fled from us, the which entered the place from which it had come, which was not of this century-world from which we had come, me and your mother Eva. The knowledge, however, entered **the seed of a great time**. For this reason, <u>**I myself have named you the name of the man who is the seed of the great generation [...] from which we come**</u>*." (Apocalypse of Adam 1:1-3) As we can see, Adam reiterates that they and their son were brought from another "aeon" (age, era, century, world, time...), and let's analyze that even the theory of pre-existence is applied throughout the Holy Scriptures to the Christ, the Messiah, affirming that he was "*promised from before the foundation of the world*." To whom was it promised if the world did not yet exist? If humans did not exist -supposedly- who was promised their coming and for what purpose? If in Christianity it is believed that Christ came to give his life for humanity, this is the same as assuming that Jesus was destined, before the world existed, to come to suffer for humanity, with which it was known that Adam would fall and that sin would enter bringing death to man. If this is true, at least in this way, God would be insensitive. Rather, if Christ was destined for the world, it was for another purpose, and to this was added, due to the present problem, the fact that he had to immolate himself. And if man was destined for the world, God being wise as he is, "something" or "someone" had to damage the plan at a certain point in history, or at least modify it, or force it to be completely redesigned.

The Big Bang

In the 9th of the Emerald Tablets of Thoth, it was written: « *Long ago in the past, I sought wisdom, knowledge unknown to man. Far in the past, I traveled into space where time began. I always sought to have knowledge to add to what I had. However, I found that only the future held the key to the wisdom that I thought.*" We have dealt with the issue of God and his environment, together with the pre-existence of the celestial man, but there is a point where that apart ends and the "fracture" of existence begins. For this reason, with a brief change of scenery, we are going to explore the universe that surrounds us, of which we are a part. Certainly, in Gnostic texts we read how Sophia, Wisdom, being a participant, and one more, of the imperishable beings, wanted to create something by herself, but without the participation of her consort and/or the other imperishable elementals. Now, that seems to have given rise to the creation that we know, not to everything, but to the structure of what exists -at least in our solar system. For this reason, before continuing to analyze the details of this problem that led to the origin of chaos, darkness and the abyss, we will analyze what science has come to know, or at least what has been officially and/or publicly exposed to us. This, as we already know, begins with the Big Bang theory, which, although apparently correct, even today its principle, the explosion, remains a mystery. The concept of the Big Bang suggests that everything that exists on the physical plane, visible and, in many ways, invisible to the human eye, resulted from a colossal primordial explosion that released the plasma that makes up the so-called known universe, throwing out fiery balls and matter that It gradually organized itself as stellar clusters in a nebular shape, many of which came to be seen in a spiral fashion or simply as extensions of multicolored gas. Although this does not answer whether "the chicken or the egg" came first, it is a principle from which official science begins.

A common question for many monotheists is: was the creation of the material world first or the rebellion of the "bad guy"? Apparently both events came almost at the same time, but the issue here is not only to think that something must have started the Duality which Adam called "eras *and competitions divided in anger* "-, but, could it all have been the product of the universal mind? What I want to say is that the studies that have been done, together with existing information about it, suggest that a thought is capable of creating, and that even physical creation is subject to the non-physical universe. In other words, everything we see seems to be, according to studies, the product of the development of a thought on a previous or higher level. A good example would be to say that we are in a dream created by ourselves; a dream that we believe to be real because we don't know any other parameter to evaluate what is real (let's remember the movie The Matrix), and that happens at a lower level than the initial shot. It seems real to us, but how can we compare it if we don't know anything else with which to weigh it? And if so, if there is an origin of everything, where is that source? Does this have anything to do with what scientists call the Unified Field? This is not a new theory, you can easily investigate it today, even by reading the studies of Dr. Groddeck, Greg Braden, or other modern scientists. For example, Albert Einstein said that *"man's experience is an optical illusion of his consciousness."* It is also known that the brain emits waves, gamma rays (34-60 Hz) which, according to recent studies supported by certain scientists, are responsible for linking neuronal time and space and interrelate in reality as a complete interpretation (memory and conscience). Hence, some say that man is the product of an interference wave generated between his electrodynamic field and the Schumann resonance. The American

Gary Renard, in his work "The Disappearance of the Universe", affirms that disciples of Jesus, specifically Thomas and Thaddeus, would have told him concepts about this illusory world created by the mind: «'wide is the road that leads to *destruction and narrow the one that leads to life', is a reiteration that he tells them how to know what life is and know it, the rest does not exist. I wasn't scaring them. That is why Jesus said: 'I have overcome the world', because he realized the illusion.* This is what some believe the Hindu prince Siddartha Gautama (one of the Buddhas) meant when he said: "*I am awake.*"

These ideas explain, in this case, that time can be altered, given that although our experiences are of linear beings, we could actually be non-linear beings. So what is time? Doubts that always bounce in our head are also the reason for these and many more constant analyzes that man performs to find answers to the "I". To go no further, some say that the soul is our unlimited uniqueness as a spirit and the real world is an intermediate between being in the body and being enlightened. But is this understood? Certainly there are so many existential questions, so many questions about the universe itself. So much time has passed since the beginning of everything and the universe is so large that other methods must be found to arrive at clear answers. The Big Bang is just one more link in the annals: «*No scientist will ever be able to trace what happened before that (the Big Bang), except on a theoretical level. But it is possible to remember the very principle and change your mind about it.*» (Gary Renard, The Disappearance of the Universe) This is totally affirmative, since there are only vague hypotheses about what was there before the Big Bang, starting with questions about how the entire potential of the universe could come out of something smaller than a neutron: not there is no scientific answer to that, only possible approaches that will never be verifiable. Nothing comes from nothing, because nothing is nothing and cannot generate something. It is evident that if God is everything, and is not subject to time, places or dimensions, and if He told us things, we could have a good and reliable eyewitness to find out what was before what is. But reality can be understood in various ways, depending entirely on the observer himself: for me, looking from below, a tree can be a tall brown trunk with green leaves, while another analyst, looking at it from above, only sees a uniform mass of green leaves, does not observe the trunk or the height of the tree.

Gary Renard also wrote in his work, from a metaphysical perspective: «*The overwhelming magnitude of painful shame and acute guilt in your mind, the result of what you think you have done; seem to require complete and immediate flight. So you join the ego and then the incomprehensible power of your mind to create illusions as a preceptor instead of creating spirit as a creator makes your method of escape manifest. At this point the ego, with which you are now fully identified, employs the ingenious if illusory method of projection to throw the thought of separation out of your mind, and you - or at least the part of you that seems to have a conscious - you seem to have been projected with him. That instantly causes what is popularly known as the Big Bang, or the creation of the universe. Now you seem to be in the universe, but you don't realize that you are literally out of your mind. Now, the enemy that you fear so much, God, no longer seems to be in the mind with you, where you think you would not have had the slightest chance before Him. Instead, God, and in this sense everything else, seems to be completely outside of you, the source of your problems, including guilt, is now somewhere else, although we have already made it clear that there can be nowhere else, the fabrication of the cosmos is your protection from God, your ingenious hiding place. At the same time, the universe itself becomes the ultimate scapegoat.*" In short, did our mind create the universe? It is evident that everything that is defined in this position is of a metaphysical order, a way of seeing things, and in part it is reminiscent of the fact that trying to think or do something outside of what is God, Above, generates something far from God. Because of our way of perceiving things, idealizing them and thinking, we hide from everything that is essentially God, separating us from the reality that is true and giving rise to duality. Something like this is a way of understanding, from this perspective, how this experience that we know as the cosmos and our place in it has existed.

Mr. Renard also added part of his conversation with one of the apostles, supposedly, on the subject, thus:

Gary: So the creation of the universe was the fourth split of the mind and that produced the Big Bang and what later seemed to be a series of almost infinite splits, or the world of multiplicity. As long as I continue to believe in the reality of this universe, then, by definition, I will also unconsciously believe that I am separate from God, and that I am guilty of filth.

Arten: Yes. You have a way with words, brother. Everything you see, from the moment you dream you are born to the moment you dream you die, and everything you dream in between, is a symbol of the one thought that you yourself have separated from God. Heaven appears to be completely broken into an infinite number of fragments and to have been replaced by its opposite. However, the history of the universe - past and future - is only a script that was written by the ego - the history composed and glorified by this idiot world - which represents, in every conceivable way, the act of separation.

» (The Disappearance of the Universe) This philosophy seems to want to teach that the current existence is something fleeting to which we give too much importance, as if it were the only thing. He wants to expose that we really are universal creatures tied to concepts that, honestly, do not stop being tics to our own mind. The primordial point of this approach arises from the relevant aspect of the separation from oneness with God, which is presented in another way in the writings of Nag Hammadi – despite the fact that they continue to imply that everything has to do with the separation of the indivisible unity with the All, provoking something foreign to that luminous perfection. However, another part of said interview mentions:

Gary: That reminds me of what we call the Big Bang, which is the symbol of the separation and projection of the universe. Does that mean that the universe will start to shrink only to eventually collapse on its own?

Arten: No. It is true that the Big Bang symbolized separation, but you should remember that, at the level of form, it was so tremendous that it produced an unimaginable amount of energy. This, in turn, predetermined all physical laws and the fate of every cell and molecule: how each would evolve and which direction it would follow. When we say that the film has already been filmed, we are saying that everything that was apparently going to happen was set in motion at that moment, and that, in fact, it could not happen in any other way. All the different dimensions and settings are simply symbols of different big bangs within the Big Bang that occurred at that very moment. And although it was all over immediately, you still have to wake up to recognize reality.

Gary: So, at this level, we suffer under the illusion that we are the ones who shape our destiny, when the truth is that the physical laws were already set in motion at that moment. Did everything that happened have to happen exactly as it did, regardless of what we tried to do about it?

Arten: Right. The mechanism by which bodily robots are manipulated by unseen forces is part of the default script representation. Remember this, since it was your decision to side with the ego and make the separation real, that means that the script is an example of self-determination, that it was agreed upon by yourself on a different level, and that you are not a victim. [...] What you observe here is like a recording of the ego, and it is up to you to decide to listen to another tune. The universe doesn't have to collapse in on itself to end. What has to happen is the collective awakening; then the dream universe will simply disappear, because it was never anything more than a meaningless dream. To keep you engrossed, the script seems to go faster and faster as time goes by, your attention span getting shorter and shorter. Until you destroy your civilization and start over again with very little memory... as if you started a new life. » (Gary Renard, The Disappearance of the Universe)

Science commonly affirms that the Big Bang was a great matter that arose from nothing and that it is still moving, and will continue to do so, although there are also those who affirm that it will actually reach a point where it will implode, having returned to the starting point through of a retrograde effect. Mariam, one of Jesus' disciples, a native of the city of Magdala, once asked Jesus, surely after the Resurrection of the Lord: « *So, will matter be destroyed or not? The Savior said: "All natures, all productions and all creatures are implicated among themselves, and they will dissolve again in their own root, since the nature of matter dissolves in what belongs only to its nature. Whoever has ears to listen, let him listen."* (Mary Magdalene 1:7) We will clarify this point throughout this book, since we have to clarify the unknowns in relation to the deity, since certainly the entire universe is unified, including God and his creatures, being us the ones who see you apart Although, that explosive beginning brought a law to which everything under it is subject: space and time. While the eternal and superior does not submit to anything, the created is subject to what is Above and is limited by many forces, including this space-time: «Then *HE, the Master, spoke: Know yourself, O Thoth, in the beginning there is VACUUM and nothingness, an eternal nothingness, without space. And within nothingness came a thought, determined, all-dominant, and filled the VOID. There was no matter there, only force, a movement, a vortex or vibration of determined thought that filled the*

VOID. And I asked the Master saying: Was this thought eternal? And the DWELLER answered me: In the beginning, there was eternal thought, and for thought to be eternal, time must exist. So within all dominant thought grew the LAW of TIME. Yes, the time that exists through all space, floating in a smooth rhythmic movement that is eternally in a state of fixation. Time does not change, but all things change in time. For time is the force that keeps events apart, each in its own proper place. Time is not in motion, but you do move through time as your consciousness moves from one event to another. » (Table 10. Emerald Tablets of Thoth)

the multiverse

Some authors of UFO themes such as Sixto Paz Wells, Alex Collier or the Urantia Book itself, cite matters from the Mental Universe, although it is time to study each one to see what is true from there. The Duality would have taken place at the moment that the collective mind of said Mental Universe wanted to think about: what would it be if...? Or what if...? Basically what it would be like to be separated from God, as we see what happened with Sofia (it is possible that the event of Sofia personified said action). At that moment, that mere thought, with the capacity of the power of the mind, created the multiverse or existing creation, which I do not know if it is really personified in Sofia, since she, although it is an identity, in reality, is not something outside of... or separated from... since in God we are all one and the same -more than anything at that moment where consciousness was unified. Certain ufological positions affirm that there are at least 22 parallel universes and in our universe 11 dimensions, while we are in the Third Dimension, on the way to the Fourth Dimension. The five theories of Super Strings, based on the concept of energy, also lead to think about ideas of this caliber that began to have more space when the Amanda telescope discovered more dimensions than the three known. However, many approaches to String Theory or Superstrings seem outdated and totally out of context, going so far as to postulate that there are multiple "I"s living in various universes at the same time, something that personally, for me and what I have been researching for some time It doesn't make logic. In any case, many other things do seem to make sense: «*Know, oh man, that all space is filled with worlds within worlds; yes, one within the other, however separated by Law.*» (Table 10. Emerald Tablets of Thoth)

Ergo, the word "Multiverse" means "Multiple Universes", that is, "Multiple Verses" or "Multiple Versions". For example, in the theories of the Urantia Book it is said that the Earth (called Urantia) is in the universe called Nebadon, which would be the Local Universe; this Nebadon would be accompanied by other creations, and all would be included in a "super-universe" called Orvonton, whose capital would be Uversa. The point is that, already in the field of revealed documentation, there are curious and striking chapters in this book, but others that go beyond the proper sense of historical machinery and contradict the Scriptures and scientific discoveries. Now, in the sense of creation, still in development, these hypotheses formulated there say that this super-universe, Orvorton, would be one of 7 existing ones. It is said, according to this theory, that these Seven Super-Universes would make up the Grand Universe or part of the Master Universe. In this case, the capital of this Grand Universe would be a central divine universe called Havona. This Havona would come to be a type of stationary "island" of Paradise or geographic center of infinity or the dwelling place of the eternal God. So, this Grand Universe, made up of said 7 super-universes, would be only a portion of the Master Universe, full of other universes from outer space, many of them uninhabited but in the process of mobilization. Although I reject many of the claims in this book, I accept that the way some things are posed suggests thinking about how they really are, giving plausible insights into the way the universe develops and is organized. Since the book itself constitutes a sum of various different stories, from different sources, it is conceivable that there are true things and false things, being, in any case, an aid to have a better perspective, but not an entirely reliable source, since it seems disinformation based on certain certain things, especially with solid New Age undertones.

Since the term "galaxy" is very short, it might be suggested in The Urantia Book that they use the definition "universe" to refer to it, that is, to name a living galaxy. When I say living galaxy I mean galaxies that are in full development of life and with organized management systems. « *The relatively quiet zones between the levels of space, such as the one that separates the seven superuniverses from the first level of outer space, are huge elliptical regions of space activities. These zones separate the vast galaxies that speed around Paradise in orderly procession. You can conceive of the first level of outer space, where incalculable universes are now in the process of formation, as a vast procession of galaxies revolving around Paradise, bounded above and below by the rest zones of space between, and bounded at the inner and outer margins by relatively quiet zones of space."* (Chapter "The Spatial Functions of Paradise." Urantia Book) This theory would then suggest that our solar system fits the Nebadon concept, Orvorton fits the Milky Way, and what they call the Grand Universe is the Local Group. of neighboring galaxies. It is said, in other UFO theories, that we are 7 or 9 galaxies that confirm the local group and one of them is in the center contemplating how the rest revolve around it. In the Urantia Book theory, this would be Havona, and for us it would be the galaxy M-31, mythically called Andromeda. Could this hypothesis fit with logic? The point is that in any case the theory of the Multiverse as well as Super-String Theory is suggestive, and more and more space is opening up in science. The Urantia Book has many and great links and inconsistencies -according to the opinion of this researcher- but it helps to open perspective and vision, being able to have the root of its data in something real, but aimed at cementing the ideas of the New Era (psychological engine of society to help consolidate ideas that prepare them for the arrival of the Antichrist).

Science today has knowledge of other dimensions and this was land that ancient civilizations had already entered. Just as there is knowledge about the discovery of the Ley Lines and the Star Gates, so there are also records and testimonies of their use. The portals would have been used by the Egyptians, the Sumerians and the Atlanteans, to enter other realities and connect with other beings, or simply open a place for them to access our plane. None of this is fiction and it also seems to be in the process of being repeated, reusing these portals again, such as the one that seems to be in an international conflict in the Gulf of Aden, since 2009, approx. It cannot be said that it is by chance that contact groups mention similar things, that is, that there are «*six spiritual planes of vibration, positively, starting with level 1, which is that of incarnated beings. Spiritual levels 2 and 3 follow, which are planes of error. Plane 4, which is a Mastery level, and Plane 5, where the spirits of maximum Light dwell. The 6th plane belongs to the angelic world. Actually, it is not above the 5th plane, but giving an earthly example: a dolphin and an elephant live on the same plane, but in different habitats. In addition, there are two levels below the physical plane: level -1, which is a vibration where the spirit is completely isolated, and level -2, where each spirit shares the pain of all the other spirits that inhabit that plane. Besides, there is the 7th vibrational level, where the Divine Energies live and the 8th vibrational level, of the Elohim or minor gods. The "spiritual and angelic" world consists of a physical plane, five positive spiritual planes, and two negative planes. They are represented in the form of surrounding spheres, where the largest contains the next and so on.»* (Message from Contact groups) Disconcerting, but will we believe everything they say that it comes from extraterrestrials? You have to study everything, because the Nephilim also use these "channels" and "mediums" to deceive, even in the world of Christian churches.

The Peruvian researcher Sixto Paz Wells (RAMA Project) affirmed something similar many years before, and continues to maintain it in the present: « *We live in a 7-dimensional Material Universe, above these there is a 3-dimensional Mental Universe and above this, a Spiritual Universe. The Spiritual Universe would have created the Mental Universe and the Mental Universe would in turn create the Material Universe.*" In Meno's dialogue, Plato would have spoken about the Material World and the World of Ideas; This theory consists of accepting together with the physical world, constituted by material, sensitive, particular, mutable, compound, generable and corruptible bodies, the existence of a world of Ideas or immaterial, intelligible, universal, immutable, indivisible and eternal forms. It is said that knowledge of ideas is science, while knowledge of things is just opinion. Plato believed that ideas are the essence and cause of all things, they are what we think of as concepts and that we designate with a name. Ideas really exist in a world apart, outside of things and the human mind; they are the authentic reality in front of the apparent sensible reality. Each of these ideas takes us back to the starting point: Divine Thought. What is thought capable of doing? The Asian scientist Masaru Emoto already demonstrated years ago how human thought or an image can transform microscopic forms of water. This has also been studied with plants and animals. We can influence our neighbor for good or bad, just with a facial expression, with a gesture, with a look... and with a thought? Everything that we create or do before, previously, we capture in our mind, that is, we think about it -or at least that's how it should be: think before acting-, because it is the way the universe moves. That is why thought is energy, and it can be perceived there as it can influence, even create, since art is the very embodiment of ideas and imagination.

From a surprisingly remote text, found in Mesoamerica, we find: «*When you are free from the shackles of darkness and travel in space as the SUN of LIGHT, then you will know that space is not limitless but truly limited by angles and angles. curves. Know, oh man, that all that exists is but an aspect of greater things yet to come. Matter is fluid and flows like a stream, constantly changing from one thing to another.*" (Table 9. Emerald Tablets of Thoth) The Amanda telescope discovered the existence of additional dimensions, that is, more than the 3 known ones. Other scientists already reached these statements considering the existence of between 7 and 10 dimensions. The ancient writing of the Yucatan, which I have referred to, says: « *9 are the interconnected dimensions, and 9 are the cycles of space. 9 are the diffusions of consciousness, and 9 are the worlds within worlds. Yes, 9 are the Lords of the cycles that come from above and below. Space is filled with hidden things, since space is divided by time.*" (Table 9. Emerald Tablets of Thoth) Contact groups are not exempt from receiving this type of information about universes and dimensions, although, as I said before, it is necessary to know how to discern the origin of said knowledge, since it can come from "evil". how it could come from "good" (it is up to one to study this). They also speak of levels or dimensional planes within each world. In other words, not a planet, but a world... observe this distinction carefully. We use the word world to describe our planet Earth, but the original source of this term goes much further. The Greek word "cósmon", from which Cosmos comes, designates the "universe of life", "universe of the living" or "universe that has life"; the same as the word in Hebrew "olam". Why do we give it such limited use? Jesus said: «*...You are from below, I am from above; you are of this world, I am not of this world.*"

(John 8:23) What world was Jesus from? Because he repeated the same thing to the Roman procurator Pontius Pilate: «*Jesus answered: 'My kingdom is not of this world; if my kingdom were of this world, my servants would fight so that I would not be handed over to the Jews; but my kingdom is not from here'.*» (John 18:36) A world is a life system, since there are planets that do not present the conditions to be a "world" -despite the fact that with advanced technology colonies can settle on it or within it. Hence one says that someone "is in their own world" or that each person "is a world."

If it were the case that all this created universe actually came to be an illusion, this would explain the scientific results of Quantum Mechanics and Quantum Physics, that is, the study of the "Quantum". The Hannover detector, the GEO 600, could have proven that we indeed live in a three-dimensional hologram: « *The GEO 600 gravitational wave detector, in Hannover (Germany), recorded a strange background noise that has brought researchers to their heads who work in it. The current director of the US Fermilab, the physicist Carl Hogan, has proposed a surprising explanation for this noise: it comes from the confines of the universe, from the corner where it goes from being a smooth space-time continuum to being an edge. granulated. If this theory is true, said noise would be the first empirical proof that we live in a holographic universe, says Hogan...*» (Reporting by Yaiza Martínez, Scientific Trends) Alex Collier, an American who claims to have been contacted by extraterrestrial humans from the Andromeda galaxy, says they let him know that our universe is like a 21 billion-year-old hologram. This statement fits with others from sources other than ufology. And with that? That if all this is true, or the essence of it, we can understand that the Kingdom of the Father is much more than we imagine and that there is a purpose developing from the beginning. If only the universe that we perceive has multiple dimensions, how many things can be there in planes that our eyes do not see? Light itself, doesn't it seem like an element in our universe that seems to coexist from another much deeper and pre-existing reality? It is an idea that I leave in the air. We understand that there are things that escape our eyes and we appreciate them through infrared rays. They are only invisible to the eye, but they are there. The same thing happens with radio waves... we don't feel them except with devices that capture their transmissions. If this is so with such simple things, how much more can there be that we don't know about yet? That, precisely, can lead us to the terrain that we need to analyze.

Someone would think that if the universe were a three-dimensional hologram, how is it that we feel, touch, palpate, experience and influence? It has been shown that space is not empty. Even the apparent vacuum that is almost at the temperature of Absolute Zero is, in fact, energy: electrical particles. We are energy, charged with watts and generating body heat. Otherwise, we could not receive an electric shock from a defibrillator to revive ourselves from cardiac arrest, nor be neutralized with a taxer, like those used by police officers to reduce someone from a distance by simply electrifying them. On a subatomic, all-encompassing scale, the atom itself is made up of positive, negative, and neutral particles, which we call "charges." Those are the protons, electrons and neutrons, moved and maintained by the power of God that manifests itself in the pure energy that vibrates in every corner of the universe. For this reason we can experience a hologram that is embodied by mere intertwined energy in every corner, just as we are tremendously startled by simple dreams that, sometimes, can even cause us physical pain without having had any physical contact (in this case it seems that the mind creates the reality that the body experiences to the degree of materializing the thought that has developed in the dream). Now, if there are dimensions, higher and lower, that is why the spirits are seen sometimes and not others (because they themselves have energy: ectoplasm). Hence the talk of the "world" of the dead and the dream world (of dreams). Something similar could be appreciated from the chronicles of Enoch and his fabulous and detailed revelatory journeys documented. Therefore, we are left with the question of the other universes. If there are other universes, it is possible that there is the grcat harmony of the Universal Father. How to visualize this properly? It is as if we have been taught all our lives that there is only one room where we live, but the truth is that there are many more rooms, hundreds of them, making up a gigantic building. Then, when we became aware of the existence of said enclosures and of this building, we observed outside of it that there are many other similar buildings with their respective rooms. With this allegory I expose that the rooms would be the equivalence to dimensions, and that the buildings would be the same as universes. The question is, how do they interact, interrelate and intertwine? Prayer itself becomes a communication bridge with higher realities. In this sense, how does this fit with the biblical and para -biblical references to heaven? Let's take into account that, depending on the specification, we speak of 3 skies, 7 skies or 10 skies.

All this existence, defined as "aeon" is referred to in ancient texts that have been discovered, which have theological writing of apparently Greek origin. In them the perception of this "aeon" is indicated, translated as "century": «- *Listen, my son, what there is of God and of Everything. God, the Century, the World, Time, the Transformation. God created the Century, the Century the World, the World the Time, the Time to the Transformation. The reality of God, so to speak, is Good, Beauty, Happiness, Wisdom; the reality of the Century is identity, that of the World is order, that of Time is change, that of Transformation is life and death. The energy of God is Intelligence and Soul, that of the Century is permanence and immortality, that of the World to go and return from the starting point to the maximum opposition, that of Time wax and wane, that of Transformation, quality. Therefore, the Century is in God, the World in the Century, Time in the World, the Transformation in Time, and this is how the Century remains stable around God, the World moves in the Century, Time passes in the World, and the transformation evolves in Time. Consequently, the source of all things is God, the reality of things is the Century, their matter is the World. The Power of God is the Century, the work of the Century is the World, which never began but is eternally engendered by the Century. Wherein the World will never perish - the Century is immortal - nor will anything in the World ever be destroyed: the World is totally surrounded by the Century. - And what is the wisdom of God? - The Good and the Beauty and Happiness and the total virtue and the Century. The Century then created the world with order and beauty, putting immortality and permanence in matter. Indeed, the generation of matter depends on the Century, as well as the Century, in turn, on God. Transformation and time are in Heaven and on Earth, but they have different natures: in Heaven without change and indestructible, on Earth with change and destruction. And God is the soul of the Century, the Century of the World, the Heaven of the Earth, and God is in intelligence, intelligence in the soul, the soul in matter. All Things Through the Century."* (The Intelligence to Hermes. Treatise XI, verse 2-4. Corpus Hermeticum)

III.

THE FIRST REBELLION

THE WISDOM

"Silence is the most powerful form of teaching that the teacher can transmit to the disciple. There are no words to express the most important things, the deepest truths."

(Raman Maharshi)

Wisdom does not find a place where it can dwell, so its house is in the Heavens. Wisdom went to dwell among the sons of men and found no place. Then Wisdom has returned to her home and has taken her seat among the angels. And injustice has come out of its caves, has found those who did not seek and has dwelt among them, like rain in the desert and like dew on the thirsty land. " (1 Enoch 42:1-3) What is Wisdom? Called in Hebrew "Chajmá" (which also translates "Cunning") and in Greek "Sophia", Wisdom seems personified, not only in all the apocryphal texts, but also in the Bible itself; referred to by Solomon as the primordial companion of the God of the Hebrews: «*Jehovah possessed me in the beginning, already of old, before his works. Eternally I had the principality, from the beginning, before the Earth. Before the abysses I was begotten; When there were no fountains abounding with water. Before the mountains were formed, before the hills, I had already been generated; He had not yet made the earth, nor the fields, nor the beginning of the dust of the world. When I formed the heavens, there I was; When he traced the circle on the face of the abyss; When he affirmed the heavens above, when he affirmed the sources of the abyss; When he put his statute to the sea, so that the waters would not transgress his commandment; When I was establishing the foundations of the earth, I was with him ordering everything, and it was his delight from day to day, having solace before him at all times. I rejoice in the habitable part of your land; And my delights are with the sons of men."* (Proverbs 8:22-31) Simeon, the son of Jesus, the son of Eleazar, the son of Sirah, mentions her in an ancient writing, saying: «*Alone I traveled the roundness of heaven, and through the depths (depths) of the abysses I walked."* (Syracid or Ecclesiastical 24:5) These words of Simeon and King Solomon lead to the writings of Nag Hammadi, where it is stated that Wisdom accompanies a deity mentioned there from the remote past, which seems to correspond to the biblical Jehovah. The words of Solomon, as well as his father David and the Edomite Job, do not stop emphasizing ancestral events of which there is no longer any memory, but which make it clear that the power of the Most High made the Earth and did so accompanied by Wisdom, not as a quality but as a deity: " *All wisdom comes from Yahweh, and is with him forever."* (Sirach 1:1)

Beginning with the proto-Hebrew prophet Enoch, it is exposed that Wisdom tried to settle in one place from the beginning. Although, having his place, he wanted to be in the world, but we humans rejected his advice, so his proper place was among the celestial ones. Thus it is understood that she has a complementary throne, next to the Lord of the Spirits, and dwells among the messengers of God. Even so, she has participated and shared her knowledge and intelligence with men who seek the truth, fear God and live in piety, according to her, rejoicing in *"the inhabited part"* of the world. Solomon reported that Jehovah *"possessed"* her, exposing the virtues and qualities of this great sovereign God, working masterfully in the creation of things, and even before this planet was designed. She, through the mouth of David's successor, points out that he had always had *"the principality"*, thus assuming that he received the secondary place after Jehovah himself and held primacy, advantage and superiority in relation to all things. In turn, she assumes that before the abysses, the earth, the sky and everything that boiled in the beginnings of this globe, she was accompanying God and being "*his delight*." In any case, as is argued, the creation of what we know arose from the determination of Wisdom: "*For those who were in the world it had been prepared by the will of our sister Sophia*" (2nd Set or 2nd Treaty of the Great Set), said an imperishable to Set, which was possibly Jesus, since he also says: «*But I am only Sophia's friend. I have been in the bosom of the Father from the beginning, in the place of the children of truth, and greatness.*" It is not at all strange, then, to see how Wisdom is personified so many times in a multitude of ancient texts, associated with Jesus as his friend, but also reflecting his maternity, possibly spiritual. Specifically, in the Book of Wisdom, it is not only defined as a "spirit friend of men", but as the Holy Spirit (Holy Spirit), the one who instructs: "*Wisdom does not enter a soul that does evil or inhabits in a body subjected to sin. Because the holy spirit, the educator, flees from falsehood, turns away from foolish reasoning, and feels rejected when injustice occurs. Wisdom is a spirit friend of men, but it will not leave the words of the blasphemer unpunished, because God is the witness of his feelings, the truthful observer of his heart, and listens to everything his tongue says.*" (Wisdom 1:4-6)

The writer of this book, according to some Solomon, adds later: «*Wisdom is luminous and never loses its brilliance: it is easily contemplated by those who love it and found by those who seek it. She anticipates making herself known to those who desire her. He who gets up early to look for her will not be tired, because he will find her sitting at his door. Meditating on it is the perfection of prudence, and whoever stays awake for its cause will soon be free of worries. Wisdom searches everywhere for those who are worthy of her, she appears to them benevolently on the roads and meets them in all their thoughts.*" (Wisdom 6:12-16) In the first century, Jesus defined this fabulous, powerful and dear friend of ours like this: «*And I will ask the Father, and he will give you another Comforter, to be with you forever: the Spirit of truth, which the world cannot receive, because it does not see him, nor does it know him; but you know him, because he dwells with you, and will be in you.*" (John 14:16-17) That Spirit, broken down in various ways and acting in a multitude of ways, takes its role from the beginning of time, but worked with the disciples of Jesus in a special way, complementing the teachings of Christ, for he has always been advising the great prophets. Jesus then went on to affirm: "*But the Comforter, the Holy Spirit, whom the Father will send in my name, he will teach you all things and remind you of everything I have said to you.*" (John 14:26) Putting it another way: "*But when the Comforter comes, whom I will send you from the Father, the Spirit of truth, who proceeds from the Father, he will bear witness about me.*" (John 15:26) So the one who would manifest Christ to the world would be precisely this spirit from God, Jesus' friend Sophia, who in turn directs the messengers who are in charge of ministering to the Lord's servants. He also expressed it this way: «*But I tell you the truth: it is to your advantage that I go away; for if I did not go away, the Comforter would not come to you; but if I go away, I will send him to you. And when he comes, he will convince the world of sin, of justice, and of judgment. Of sin, because they do not believe in me; of righteousness, because I go to the Father, and you will see me no more; and judgment, because the prince of this world has already been judged. I still have many things to tell you, but you cannot bear them now. But when the Spirit of truth comes, he will guide you into all truth; for he will not speak on his own, but whatever he hears he will speak, and he will make you know the things that will come. He will glorify me; for he will take from what is mine, and will make it known to you.*" (John 16:7-14)

King Solomon, speaking about the origin and qualities of Sophia, wrote: «*In her there is an intelligent, holy, unique, multiform, subtle, agile, perceptive spirit, without stain, diaphanous, unalterable, lover of good, sharp, free, benefactor, friend of men, firm, sure, serene, who can do everything, observes everything and penetrates all spirits: the intelligent, the pure and even the most subtle. Wisdom is more agile than any movement; because of its purity, it goes through and penetrates everything. She is an exhalation of the power of God, a pure emanation of the glory of the Almighty: therefore, nothing stained can reach her. She is the radiance of eternal light, a spotless mirror of God's activity and an image of his goodness. Although she is only one, she can do everything; remaining in itself, it renews the universe; from generation to generation, enter the holy souls, to make friends of God and prophets. Because God loves only those who live with Wisdom. She, in fact, is more radiant than the sun and surpasses all the constellations; it is more luminous than light itself, since light gives way to night, but evil does not prevail against Wisdom. She spreads her strength from one end to the other, managing everything to the best of her ability. I loved her and searched for her from my youth, I tried to take her as a wife and I fell in love with her beauty. Her intimacy with God highlights the nobility of her origin, because the Lord of all things loved her. She is initiated into the science of God and it is she who chooses her works. If wealth is a desirable commodity in life, what is richer than all-doing Wisdom? If prudence is what works, who else is the creator of everything that exists? Do you love justice? The fruit of her efforts are the virtues, because she teaches temperance and prudence, justice and strength, and nothing is more useful than this for men in life. Do you also want to have a lot of experience? She knows the past and can foresee the future, she interprets the maxims and deciphers the enigmas, she knows in advance the signs and wonders, the succession of times and times.* » (Wisdom 7:22 to 8:8)

Wisdom is also called Faith, in Hebrew "Emunah" (from the voice "Emet", which translates: "Truth") and in Greek "Pístis": «But Faith is a guide over the deep abyss and *perseverance [in] the path in the middle of the sharp rocks*." (Words of Jesus. The Sevenfold Peace, Gospel of the Essenes 1:92). These names and identifications can lead one to assume that she has been advising humanity since the beginning, and she is the one who presided over knowledge about things until Jesus came and perfected these words, speaking fully and clearly of the Father, however, also accompanied he, and many others, of herself. What's more, rarely, unlike Jesus and Solomon, someone spoke in detail and openly about Sophia, so to understand this you have to carefully analyze the teachings of millennia ago and the ancient teachers. Jesus said: «*Therefore the wisdom of God also said: I will send them prophets and apostles; and some of them they will kill and others they will persecute, so that the blood of all the prophets that has been shed since the foundation of the world may be required of this generation, from the blood of Abel to the blood of Zacharias, who died on the altar and the temple; Yes, I tell you, it will be required of this generation*." (Luke 11:49-51) Simeon (son of Jesus, son of Eleazar, son of Sirah) also spoke of this "wisdom of God" when he wrote: "*How great is the wisdom of the Lord! Strong is his power, he sees everything.*» (Sirach 15:18) It is possible that she was the adviser of Jehovah and his work through his messengers. Moreover, the Wisdom of God, is it separated from God to speak for itself? Shouldn't he rather say: "God, in his wisdom, said..."? Watch out for the data! It seems that it depends on God, it is with God or it is a quality of Jehovah, which accompanies him. «*Always seek wisdom. Don't let your body fool you. Stay on the path of the wave of Light. Reject the dark path. Know that wisdom is lasting. <u>Existing since ALL SOULS began</u>, **creating harmony by the Law that exists in the PATH**.*» (Tablet 9. Emerald Tablets of Thoth) In relation to the identity and work of Wisdom, the apostle Philip seems to share his knowledge in this regard, when he writes in relation to her as a terrestrial manifestation of the Holy Spirit: «*"Father" and "Son" are simple names; "Holy Spirit" is a compound noun. They are in fact everywhere: above, below, in the secret and in the manifest. The Holy Spirit is in the revealed, below, in the secret, above. The evil Powers are at the service of the saints, after having been blinded by the Holy Spirit so that they believe that they are serving a man, while they are operating in favor of the saints. That is why —(when) one day a disciple asked the Lord for something of the world—He said to him: "Ask your mother and she will share with you the things of others"* .» (Philip 1:34)

Which here is the mother: the Holy Spirit or the evil powers? If the former, then is the Holy Spirit our "spiritual mother"? And why does he talk about " other people's things"? Does it refer to things outside the Kingdom of God, of the spiritual? Well, continuing with the thread, Philip narrates: «*The apostles said to the disciples: "that all our offering seek salt for itself". They called [the Sophia] "salt", (because) without her no offering [is] acceptable. Sofia is sterile, [without] child(ren); that is why it is called [also] "salt". The place where those [...] in their own way [is] the Holy Spirit; [for this (?)] his children are numerous.*» (Verse 36) This seems to imply that Sophia/Wisdom is the Holy Spirit or has parallels with him/her. The children he has are those born of the spirit, that is clear, and it is understood that Jesus spoke of those who are the salt of the Earth (Matthew 5:13). Ergo, Felipe adds: «*What the father possesses belongs to the son, but while the latter is small, what is his is not entrusted to him. When he becomes a man, then the father gives him everything he owns. When those begotten by the spirit err, they err also because of it. For the same reason, the same breath kindles the fire and puts it out. One thing is "Ejamot" and another is "Ejmot". Ejamot is the quintessential Sophia, while Ejmot is the Sophia of death, the one who knows death, who is called "Sofia the little one."* (Gospel of Philip 1:37-39) There seem to be two Sophia/Wisdom: one that is the original, and another that is "the one of death", or "the little one" (surely the one referred to as "Material Wisdom"), so Jesus also speaks of "*the Wisdom of God*" and of "*the Wisdom of Afterthought*" – which could be the same, but I want to imply that he does not simply call her Sophia or Wisdom, as the only personification of the same. She, understood as the Holy Spirit, seems to be from whom Miriam conceives, not from God: "*The Holy Spirit will come upon you, and the power of the Most High will overshadow you...*" (Luke 1:35). that Ruach, while the power of the Most High covers that scene. Gabriel himself confirmed to Joseph that "*what has been engendered in her is from the Holy Spirit.*» (Matthew 1:20) A woman conceiving from a woman?

On the other hand, this Faith or Wisdom is the one that is always identified as the Holy Spirit, since its essence is, of course, Wisdom and Faith. For example, the Gospel of the Hebrews was lost, but Origen (approx. years 253/254 AD) commented: «*And if someone accepts the Gospel of the Hebrews, where the Savior himself says: "A little while ago my mother, the Holy Spirit, took me by one of my hairs and took me to the sublime mount of Tabor", he will be perplexed when considering how the Holy Spirit, engendered by the Word, can be the mother of Christ. But neither is this difficult for him to explain.*» (In Io. 2,6) And there is also another fragment in which it reads: « *If someone admits: "A little while ago my mother, the Holy Spirit, took me and took me to the sublime mount of Tabor" and what follows, can, seeing his mother in Him, say...*» (Hom. in Ier. 15,4) Jerome also commented (approx. year 419/420) in this regard: «*...But whoever reads the Song of Sing and understand that the spouse of the soul is the Word of God, and give credit to the gospel published [under the title] according to the Hebrews, which we have recently translated —in which, referring to the person of the Savior, it* *is said: "It makes My mother, the Holy Spirit, took a little from me by one of my hairs"—, he will have no qualms in saying that the Word of God proceeds from the Spirit, and that, therefore, the soul, which is the wife of the Word, has as its mother-in-law to the Holy Spirit, whose name among the Hebrews is feminine, RUAH."* (Comm. II in Mich. 7,6) And he also commented on another point: «And *by the way, in the gospel that we mentioned a little while ago, we find it written: "And it happened that, when the Lord had come up from the water, he came down all the fountain of the Holy Spirit, rested on Him, and said to him: My son, through all the prophets I was waiting for you to come and I could rest in you. For you are my rest, my firstborn Son, who reigns forever.»* (Comm. IV in Is. 11,2) If indeed Sophia is the manifestation of the Holy Spirit, she is the one who guides the followers of Christ and those who seek the truth and live according to the Most High: « Wisdom to her *children exalt, and take care of those who seek it."* (Sirach 4:11) It is she who has presented Christ to the nations: "*And after they had put to death the Messiah who was to come, and after having been killed, he would rise from the dead and manifest himself to the Gentiles." through the Holy Spirit."* (1 Nephi 10:11. Book of Mormon)

She, part of the "Spirit" of God, has been the guide of God's servants since ancient times, even though she personifies the material or lesser part of the Holy Spirit as an entity itself. Thus it was written: « *But he remembered the days of old, Moses and his people, saying: Where is he who brought them up out of the sea with the shepherd of his flock? where the one who put his holy spirit in the midst of him, the one who led them by the right hand of Moses with the arm of his glory; the one who divided the waters before them, thus making himself a perpetual name, the one who led them through the abysses, like a horse through the desert, without them stumbling? The Spirit of Jehovah shepherded them, like a beast that descends into the valley; thus you shepherded your people, to make yourself a glorious name.*» (Isaiah 63:11-14) The apostle also wrote: « *Sophia —whom they call "the sterile one"— is the mother of angels...*» (Gospel of Philip 1:55) What angels? Since there are good and bad, and there are those from Above as from Earth, and from the Imperishable Realms as from Jehovah's system of government, as these documents summarize. Those of the Imperishable Kingdom were practically already from the time of the appearance of Sophia, defined as children of Plesitea, according to the Gospel of the Egyptians -attributed to Set, the son of Adam-, but Sophia, who helped Jehovah, is the one who could give rise to the assistant angels of this one, since "*In the treasures of wisdom are the maxims of science ...*" (Sirácida 1:25) and also refer to it in this way the texts of Nag Hammadi and the Bible, who identifies it as part of the collective of God, understood as "Ruaj" (spirit, wind): «*And this shall be my covenant with them, said Jehovah: My Spirit that is upon you, and my words that I put in your mouth, shall not fail from your mouth , nor from the mouth of your children, nor from the mouth of your children's children, says the Lord, from now and forever.*>> (Isaiah 59:21) In any case, if Sophia is the mother of Jehovah's messengers, this came to be after the error of creation that arose from her, from where Chaos and the Primordial waters that so many mythologies cite appeared: «*After 5,000 years, the great luminary Elelet said: "Let someone rule over chaos and Hades." And there appeared a cloud [whose name is] material Wisdom [...she] looked into the regions [of chaos], her face being like [...in] her form [...] [...] blood.*» (Gospel of the Egyptians or 1st Treatise of Seth 1:56-57) If someone was determined to rule over Chaos, this could have been the second step in the development of events that gave rise to the origin of evil, although the Father seemed to know exactly what he was doing, and that's why he allowed it.

But then, is Jesus the son of the Holy Spirit, that is, of Sophia, Wisdom? The Apostle Philip wrote that "*Some say that Mary conceived by the Holy Spirit: they are wrong, they do not know what they are saying. When has a woman ever conceived of a woman? Mary is the virgin whom no Power has stained. She is a great anathema for the Jews, who are the apostles and the apostolics. This virgin that no Power has violated, [...while] the Powers became contaminated."* (Philip 1:17) Is he referring to the terrestrial Mary or is there another that is not terrestrial, and Felipe speaks of this? In any case, it is strange that on the one hand it is said that he is the son of Sofia and on the other hand that he is not, perhaps because this could lead to misunderstandings: Sofia was a bridge to bring Christ from higher realities, while the Father simply overshadowed the event. Well, Wisdom is one of the celestial creatures that were in the Undying Realms (she's even representative of a whole Realm of Immortals), but, according to the manuscripts, she wanted to create something without the joint work of the other immortals, ago long ago: «*Now Sophia, who is the Wisdom of the Afterthought and who represents an eternal kingdom, conceived a thought. She had this idea herself, and the invisible Spirit and the Former Thought were also reflected in her. She wanted to give birth to a being like herself without the Spirit's permission (the Spirit had not given approval), without her lover, and without her consideration. His partner did not approve; she found no one who agreed with her; and she considered this question without the permission of the Spirit or any knowledge of what she had decided. However, she gave birth to a Kindred. And, due to the unconquerable power within her, her thought was not an idle thought. Rather, something came out of her that was imperfect and in appearance different from her, for she had produced it without her lover. It did not look like its Mother, and it had a different shape."* (Secret Book of John 6:1-5).

But there are more references that express this chaotic beginning: «*When the unlimited character of the immortals ended, a Faith (Truth) came into existence that is called Wisdom. This form felt the desire to create a work like the light that exists from the beginning. Wisdom's desire was spread out as soon as the sky, to a size that cannot be conceived, situated in the center between the immortals and those who came after, above where there is a veil that separates the upper part of human realities. However, in the aeon of truth there is no shadow because there is immeasurable light through it. [On] its exterior, however, shadow appeared, for which reason it was called "darkness." During the power came darkness. This shadow, the power, entered after the so-called "unlimited chaos". The latter, gave birth to the race of the gods with all the place, as [...] also [...] which follows the first work of [which is] self-evident. The abyss exists [because] from the Faith [of which] we have spoken."* (Apocryphon of Coptic Creation 98:11 to 99:2) Depending on the translation, it is said that the result was "darkness" or "darkness" (this is the corresponding word according to the genesis story), and "limitless chaos" It is also mistranslated in the said book of Moses as "without form and void". That chaos reflected an abyss and also thanks to it appeared *"the race of the gods"*, something forbidden to mention in the Hebrew culture. The recklessness of Wisdom/Sophia generated something of great magnitude, and that other consequences, and so on. She is the one whom Jesus refers to as belonging to the Afterthought (which has a Kingdom in the domain of Oroiel), but seems to be identified with the Wisdom of the Kingdom of Elelet, which is quoted again and again in proverbs and incantations: *"Before Everything was created by Wisdom, intelligent prudence from eternity."* (Sirach 1:4)

Now, after the Resurrection, Jesus also narrates relevant aspects of this matter to the group of followers: «*Mary then said to him: "Holy Lord, where do your disciples come from, where are they headed, what should they do here?" The perfect Savior answered her: "You should know that Sofia, the Mother of the Universe and the wall, wanted to bring them into existence alone, without the participation of a male spouse. But, by the will of the Father of the Universe, so that his unimaginable goodness could be revealed, he created a curtain between the immortals and those who came after them, so that their consequence could follow each aeon and chaos, so that the woman could live despite its imperfections, that it can exist even though Error fights it. That curtain is the spirit."* (Sheneset Gnostic 1:20) This gave rise to the existing world, not specifying whether it is this planet, the solar system, the galaxy, or the cosmos itself. It is necessary to understand that Sofia's intention was to continue with the procedure of creating celestial identities, and man is a reflection of it, but in a subsequent aspect and perfecting Sofia's mistake and doing it with the approval of the Assembly. This event could be interpreted as the beginning of evil or error, but that aspect of Duality that departed from here only showed God's interest in the human race, Sophia's original dream and desire, descending into this reality, only which, in this case, would suppress the failure, although seen in another way, it would be the opportunity for celestial humans to experience a mortal life and start from scratch in an apprenticeship full of multiple new experiences. Seeing it another way, on a metaphysical level, it would have expressed the collective desire of the divine man, in Sophia, to separate from the oneness of God and appreciate the existence of a life of vulnerability and disability, his path being the return to the point of origin. as a return to immortality and perfection. So, did evil come from good? The answer is neither yes nor no, but both: yes and no. Good cannot engender evil, but things, arranged in a way that is not in accordance with what is perfect, generate "not perfect" things, such as the elements, which in themselves are complete, but when mixed with different properties they decompose quickly.

The Evil

Of writing whose language has been lost today, there are still vestiges of what the masters of the ages before the last era of men knew. Many eons have passed, disappearing peoples and entire civilizations, of which nothing remains, not even a narrated reference. It is not uncommon that novels survive in certain places today about times of the past in which there were magical and supernatural beings, majestic sites and also horrendous battles against dark hosts that exceeded human comprehension: «Listen, oh man, to the wisdom of magic. *Hear the knowledge of the forgotten powers. Long ago, in the days of the first man, the war began between the dark and the light. Man, then as now, was filled with both darkness and light; and while in some of the darkness hell held sway, in another part light filled the soul. Yes, ancient is this war, the eternal struggle between darkness and light. Fiercely it is fought through all the ages, using strange powers hidden from man. Adepts have been filled with blackness, always fighting against the light; but there are others who, filled with radiance, have always conquered the darkness of the night. Wherever they are in all eras and planes, surely, you know of the battle with the night. Many ages ago, the Morning SUNS descending, found the world filled with night, there in that past, the fight began, the ancient Battle of Darkness and Light."* (Table 6. Emerald Tablets of Thoth) These tablets are supposed to be around 38,000 years old, and are attributed to a king-priest of the submerged Atlantis, the same one who would have founded a colony in distant Egypt after the submersion of his mother. homeland. If these words were true, this priest would confess that already in that age, before the Deluge itself, the struggle between light and darkness was perennial. Those "morning suns" he mentioned are reminiscent of meritorious memorial stories from his own punctuated narrative, but such chronological reference places the conflict against evil in other ages even older than the one in which Thoth would have appeared. It is understood, then, that when the children of light descended to Earth, evil was already in full swing and had strong and extensive forces among its ranks. According to the sumerologist Zecharia Sitchin, a little more than 475,000 years ago (others say that it was a maximum of 275,000 years) the Nephilim (Fallen) came and, if they were related to the morning suns described by Thoth, the appearance of the rest could have been very above, implying that we are talking, not about millennia, but even millions of years.

The record that we can still scrutinize shows us that before there was light or darkness, something chaotic appeared. Certainly the majesty of the Most High and his imperishable creatures were already there, but this the world did not come to know. What remained as a written and verbal record in the peoples of the past was from the appearance of darkness and water. In Egypt the truth on the matter was collected, for this country was once the cradle of immortal beings and armies of heaven. This nation, as a testimony, kept all the mysteries of Creation, which is why it became the target of demons and usurping gods; since if they managed to dominate this region, the nations would not find out their secrets. From there come stories of the legendary gods, but the native men themselves had no information regarding what came before, believing that those gods came first – except for a privileged few. When Egypt became a major kingdom, under Pharaonic rule, the descendants of Abraham's lineage possibly carried their historical records, leaving copies in this place. It is at the same time plausible that this material was lost by the hands of the Israelites in subsequent generations, but somehow, thanks to the divine will, it was recorded hidden or passed from father to son. It is significant that first-century Christianity had the opportunity to possess this material and study it, leaving a good safeguard in a town called Nag Hammadi, where in 1945 a collection of Gnostic texts was found. That year, 13 leather-bound papyrus codices buried in sealed jars were found by local peasants. These manuscripts include 52 treatises that are usually defined as "gnostic" (from the Greek word "gnosis", which means "knowledge"), but also three works belonging to the Corpus Hermeticum -which is usually attributed to Hermes Trismegistus, Tat and/or Asclepius. - and a partial translation or alteration of Plato's The Republic. The codices are believed to form a library hidden by the monks of the nearby monastery of San Pacomio when possession of these writings was declared a heresy by the ruling empire. The codices are written in Coptic, although the works were probably translations from the Greek, given the typical common words they used, which leads one to suppose that under the auspices of the Greek powers, as far as the language is concerned, the writings were passed to koine.

Starting from this we can advance in the research work, since these texts, such as mythologies, the Bible, the apocrypha, the deuterocanonicals, the Qumran manuscripts, the Judaic material (Talmud and Pseudepigraphics) and many more sources will help us to find the hidden secrets of the origin of man, the world, heaven and evil. The differences between one and the other are only normal alterations that take place due to changes in the language, geographical displacements, transcriptions of deteriorated previous material, loss of phrases, deliberate destruction, a long time elapsed, lack of scribes and ignorance about languages that were being forgotten. Even with everything, the work of decades, collecting data and writings, thanks to the advance in technology and electronic facilities, lead to being able to find knowledge that was not possible in the past. This includes the work of understanding ancient texts and their hidden stories. As I have said, the stories of the past always refer to the same thing, but with different words, depending on the region and the migratory and war history. In the case of the Hellenes, they had the myth of Pandora, the first woman, according to Hesiod, created by order of Zeus, to punish men for the disobedience of Prometheus – who is said to have stolen fire from the gods to give it to the humans. This version says that Zeus created Pandora with the help of Athena, and gave her to Epimetheus, brother of Prometheus (whom some associate with the Egyptian Nefertari) as a wife. She, having conceived from him, gave birth to Pyrrha who would be the wife of Deucalion –son of Prometheus-, the hero who escaped the Deluge. The Mythological Library (I, VII, 2) mentions that Deucalion and Pyrrha are considered by myth as ancestors of most of the peoples of Greece, although the Bible assumes this value to the descendants of Jafet. Curious similarities with the Bible, but in this case there would be something deeper, since the legend indicates that all the gods adorned Pandora with their graces except Hermes, who made her curious and deceitful, supposedly by design of the father of the gods. In this way, Pandora's curiosity led her to open the trunk where all the evils were deposited. That is why it is said that Pandora is the cause of evils on Earth, since she opened the vault of the dark source, and when she wanted to react and close it again, only hope remained inside -because she did not have time to come out-. Although there is talk of a "box" it was really a kind of urn, jar or amphora, which Zeus would have deliberately left in front of Pandora to cause evil to the world, knowing that she would open it.

It only remains to accept that the beginning known to man was in relation to darkness, a personified force of something negative, an amazing power that was under a body of water of unspecified proportions. Apparently, that was something extremely chaotic and inconsistent with higher parameters. Those "superior parameters" are identified by one now, who knows about the Kingdom of the Father, but back then that was the basics; there was nothing strange in living matter coming out of a darkness and formless body. That black force drew attention by shouting that it was a demiurge, since it had no idea where it had come from, "*And the Spirit was frightened by the noise. He got up from his post. And he saw a water of great darkness. And it made him nauseous. And the idea that the Spirit looked, saw the infinite light. But he missed the rotten root. And by the will of the great Light the dark waters parted. And the darkness drew near shrouded in vile ignorance, and (this was) in order that the mind may be separated from him, for he [was] proud of himself in it.*" (Paraphrase of Shem 1:20-35) That part of the "parting" of the waters is notable in Moses' account, but here he identifies the bad part that wanted to be removed, in a process that at the same time contributed to the formation of a water balloon and a few others around it, by that Spirit, which is none other than Wisdom personified (Sophia). The objective of this exercise was to take the dark part and bring it under said aqueous sphere and isolate it completely, taking at the same time the creative power that was swarming there and bringing it back to its proper place. That negative force was amazed to see that Spirit of Light that emerged and separated the waters, so he himself went to the place prescribed below (what we define as Hades) and tried to create powerful forms similar to the light that had appeared (which is reminiscent of the goddess Tian-Mu of the Taoist tradition, whose name means "mother of luminosity"), but it was of little use to him, since he could only have an idea of an image, not of the power itself that inhabited that spirit of light that we call the Holy Spirit or Wisdom. That "Mind" is also spoken of as if it referred to a thought power that made decisions for itself, and that mind had to be in the right place, which, in any case, was not with the darkness.

In Persian culture, a name was known that identified evil: Angra Mainyu, also called Avestan, Ahriman or Ariman (Farsi اهریمن). This, in the classical sense, would be the evil mind of the human. Although it does not appear in the old Persian inscriptions, in the Avesta it is called "twin brother of the Holy Spirit", and the "opposite of Spenta Mainyu". He is considered the destructive

Satan, the source of all evil in the world and, like Ahura Mazda, he existed since its creation. Ahriman consciously chose evil, created disease to hasten the arrival of death. It is considered that his greatest evil was to corrupt the pure fire created by Ahura Mazda, to which he gave color and added smoke, giving it its polluting characteristic. On Judgment Day he will be destroyed by Spenta Mainyu and will disappear from the world forever. Although an analogy with Christian beliefs can be observed through the concept of Satan, it is not the same since Angra Mainyu is not considered a being or an entity. Angra Mainyu refers to the bad mentality of the human being, he is misinterpreted as the main agent of evil, despite the fact that it is accepted that he represents the daeva, those of bad conduct. Through the balance between opposites that Angra Mainyu supposes in front of Spenta Mainyu, the prophet Zoroaster distinguished the two poles of a particular dynamic: creation and destruction, contemplated as a whole in Ahura Mazda. In another example, according to Guanche beliefs, Guayota, the demon king of evil, lived inside the Teide volcano (hell). According to popular legend, Guayota kidnapped the god Magec (god of light and sun), and took him with him into the interior of Teide, plunging the whole world into darkness. The Guanches asked Achamán, their supreme god, for mercy. After a fierce fight, Achamán managed to defeat Guayota, remove Magec from the bowels of Echeyde (name of the mountain that is currently known as Teide) and plug the crater.

So, we have quite a few stories that affirm that at the beginning of creation, a being rebelled against the celestial designs and divided the cosmos. The followers of Josep Smith, for example - and not based on the Book of Mormon but on later philosophies without the support of sacred texts - affirm that in the beginning, on the planet Colhan, a supreme deity called Elohim had two main sons known as Jesus and Lucifer, and this second rebelled because he wanted to force humans to become gods while Jesus proposed giving them free will. This legend is repeated in different ways and in a similar way at some points in other ancient narratives. The Freemasons, for their part, say that in the beginning there were two deities: Adonai and Iblis. Iblis (an Arabic name to refer to Satan) rebelled and wanted glory for himself, he wanted to become the liberator of men, just as the Sumerian tablets affirm about Enki. Certainly, in that legendary Mesopotamian version, Enki would be the brother of Enlil, both sons of the supreme deity An (Anu), who ruled over that fertile land. His father, An, ruled the heavens (solar system) from his throne on a planet called Nibiru. The similarities are appreciable. Other novels suggest that Lucifer was a great sovereign leader from a planet called Satania, who announced demands for greater jurisdiction and attempted to displace all higher orders of sonship in the "local universe" governmental plan. So suggests the Urantia Book, which considers it the first rebel. In another order, we have more extraterrestrial theories that say that there was a first rebel named Helel or Lucifer, who was a being from the Mental Universe, who poisoned the mind of a general and politician from the Orion constellation, named Satanael. As it is said, both practically became a single spirit or thought because they acted in a type of psychic, spiritual and astral symbiosis.

In more descriptions it is stated that the figure of Satan and Lucifer have always been linked to the extraterrestrial beings popularly called Reptilians (called "Draconians" in ufology, according to their apparent origin: the constellation of the Dragon). Other legends, which later took root in the most proliferating monotheistic religions, claimed that Satan was an angel who became a red-skinned demon, having been banished to the underworld, from where he wished to mislead men by sending horrible, incorporeal beings to possessing physical bodies –these ideas that have been popularized in Catholicism without sticking to the Bible. In another area, the narrative about Faith (Pistis), also called Wisdom (Sophia), and its idea of creating a being without the support of the Creator, is not new. We see that she, given the great power inherent in her being, created something that was not in vain, but also gave rise to what is known as "dark waters." Strangely, those waters, divided (its center would later be an abyss), are a matter of perspective. In the same way is the "abyss", which is somewhat difficult to locate geographically, for which some believe that it is a concept that describes the spiritual fall or the lowest point of worldliness, while others associate it with the Hades. Indeed, it sometimes makes us think that he is referring to another dimension, like where the chaos would come from, or simply the darkness that overflows inside the Earth. The narratives tell us that Sakla (Saclas), Samael or Jaldabaoz (Yaldabaoth), son of Pistis Sofía (Faith Wisdom) was involved with the first creations, not because they were given to him but because he was also a great creator and artificer. Apparently, for a long time, in a universe of harmony, it was not determined what to do with it. He did what he wanted and prospered, however, his own creations began to rebel against him and the same creation was taking place. We find this related to what we could call the First Aeon and the First Creation; and later, for the Second Aeon, from my perspective, it was when another character enters the scene and unleashes a large-scale insurrection. This second individual, a kerub, possibly the one called Satanael or Jeilel Ben Ha-Shajar (Son of Dawn Force) was the one who followed in Sakla's footsteps and commanded his army: The Dragon. Now the Holy Spirit had to work hard to restore what had been stolen and neutralize this absurd diabolical cause.

Barnabas wrote: « *There are two paths of doctrine and power, the path of light and the path of darkness. Now, great is the difference between the two paths. For on the one are the angels of God, bearers of light; on the other, the angels of Satan. And the one is Lord from the centuries and to the centuries; the other is the prince of the present age of iniquity."* (Epistle of Barnabas 18:1-2) It is notorious that Satan has identified evil for a long time, but he was not the engine of negative things. Therefore, what is evil? Who is the father of evil?: «*But afterwards, [when he] is put to shame, he will be puzzled, [by the fact] that his effort is far [of] aeons, it is a nothing. And his inheritance is going to be small, that is, the one he boasts about [saying] it was great. His gifts are also not beneficial. Your promises are bad advice. For you are not [included] in his mercies. Rather, he exercises violence against you. He intends to do us injustice. And he is going to reign for some time, fixed for him.*» (Second Apocalypse of James 1:53) The concept of evil is born with the dualistic idea that from the beginning there was a good and a bad polarity, complementing each other. Thinking like this is natural in the vast majority of peoples of antiquity, except in Hebrew-Jewish monotheism, Christianity, Islam, Zoroastrianism and other small aspects. Just as in distant Persia the source of evil, from Mazdeism or Zoroastrianism, was known as Angra Mainyu, in the Hindu tradition the evil Vala was spoken of. This destructive and malevolent being received a great defeat, according to the Vedic texts, by the work of the only true god, the Creator Father, Brahma, before the origin of the gods and demons on Earth. This story has been compiled everywhere, from Mesoamerica, Australia, Egypt, China or Iceland. After the origins of evil, those who have raised and maintained the fight against God and against man have emerged. Hence, for example, Shaytan, as he is called in Islam, has been the "opponent" and "seducer" of humanity, dedicated exclusively to leading men astray: "he is a declared enemy for us." (Muhammad). Although one character is older than the other, but the older one, the previous one, is inactive, while the later one is active.

In certain parts of Africa the personification of evil was called Angat, and it is a principle represented in the form of a reptile, to whom human flesh was offered (the appearance of the reptile god or being a dragon has a primal appearance in practically all the towns of yesterday). But what is its origin? It is confusing to try to find its origin by synthesizing ancient history, since it alone is incomplete and does not speak of before the chaos. Some see the evil beginning in the identity referred to in Ezekiel 28 where it speaks of the "king of Tyre" and below the "prince of Tyre", but the bible does not deal with this matter entirely, as there seem to be missing books that have been lost. In the written history of Israel and the children of Ishmael, when speaking of Jehovah or Allah, a singular god was not cited but rather a plural form: "we did", "we said" or "we thought". In fact, if Jehovah were a single person, he would not have said: "*The Lord rebukes you*" (Zechariah 3:2), as if he were speaking in the third person. This example also serves to explain why the personification of evil seems to take different forms: it has many faces. Each of the manifestations or "children" of Evil are understood as evil itself in different times and places. This occurs in the same way when the children of light are referred to as a unit, and each and every one represents the Good; and in the army of Jehovah, all, or one, represent Jehovah. So this absolute evil or "Dark Side", also personified in JRR Tolkien's novel as "Sauron", and expressed as an "eye", is precisely the characteristic symbol of the Egyptian creator god Ra, who passed on to Horus. The Satanist and occult mentor of Adolf Hitler, Aleister Crowley – who is believed to have also been a lover of George H. Bush's mother – calls Satan "the eye", and also in Freemason and Illuminati occultism the characteristic symbol is "the eye". all-seeing eye", and therefore, the manifestation of Horus, which is expected to be seen again incarnated in a human body. Obviously the Egyptian gods did not teach that they were evil or servants of Satan, but rather that what was bad for them, like Seth, the brother of Osiris, they described as evil or as having a negative and destructive connotation, as happened in Scandinavia among Odin and the evil sorcerer Loki.

The advent of Horus (the Distant), the falcon, to incarnate in a human body, is the cornerstone of occultism and a synonym for the Antichrist, the god who will rule the world for a short time. It is this reason one of which associates the Antichrist with a one-eyed man, in the likeness of Odin (the original figure of Santa Claus), in Islam and in the occult. It is well known in Egyptology that Ra ceded his power to Horus, and his "name", since his successor, Osiris, was not fit to reign, thus leaving his miraculous first-born (Osiris and Isis's), Horus, under the leadership of this North African nation. The story of the Nag Hammadi texts about the secret power of Sakla and that of the deliberate poisoning of Ra by the goddess Isis are very similar: Ra had a power of creation and manifested itself in different ways, although he never revealed his true identity. nor did she reveal her true name, except to Isis, in secret, so that she could cure him of the bite of a serpent that she herself secretly sent to her, and that name she had only to make known to Horus, the sovereign and greatest god of all Egyptian history –even superior to Amun-Ra. The physical body created by Ra as an avatar would deteriorate and he would return to control the passage of the sun and the underworld, his great-grandson Horus being -according to one of the versions- the one who would represent him. Also in the texts of Nag Hammadi (Egypt) they say that the second assistant angel or generating spirit of the Earth, created by Sakla was Harmas, called the "jealous eye" or the "eye of fire". This is one of the "twelve angels/gods" of Sakla: 7 heavenly heads and 5 underworld heads. remarkable similarities.

We then have that the personification of evil took several names in the past, among them "Sakla" (erroneous), then

"Jaldabaot" (slanderous ram) and finally "Samael" (blind god). His own mother expelled him from the eternal realms, as in a parallel to the Greek story of Hera and her "androgynous" Hephaistos (Vulcan). Hera intended to have a son by herself, but the son was an amorphous and ugly being whom she threw from Olympus, being later restored by Zeus and installed as the forger of his lightning bolts and the great weapons of the gods, putting him, in addition, as a wife to the beautiful Aphrodite, which is associated with the Cananite Astoret. Sakla's offspring were the ones who presided over, and still preside over, his manifestations, and they are the faces with which he makes himself known -he never lets himself be seen personally-: Athot (sheep), Eloaios (mule), Astafaios (hyena), Yao (seven-headed serpent), Sabaot (dragon), Adonis (monkey) and Sabataios (flaming fire). His main generating spirit sent to Earth was destined for the Abyss, and this is Belial (the "lightless one"). All these names lead to confusion about the identity of Sakla. For example, the name Mastemá (hatred) can refer to him or to Belial (Beliar or Belias), and Beelzebub himself is not clear if he is Sakla, if he is Satan or if he is Belial. About this Belial it was written in the Dead Sea Texts: «*But you made Belial, the angel of enmity, for the pit, and you chose [him] with his helpers and his council to be a cause of evil and guilt. And all the spirits 'of his inheritance' are angels of destruction, they follow the laws of darkness, and these are directed to their [desires].*» (The War of the Children of Light against the Children of Darkness. Qumran Manuscripts) This is how Belial is a clear similarity to the Greek goddess of discord, Eris.

Indeed the Nag Hammadi texts also call Sakla the "First Ruler", the "Arcane", "the Beast" and the "Great Bull" – although he is sometimes defined as the "Instructor". His identification with the Beast (Behemoth) associates him with the Arabic Bahamut, a mythological huge aquatic monster that supports the Earth. This same symbol of the "sea giant" supporting the Earth, is seen in the Greek tradition, where Atlas holds the terrestrial sphere on his shoulders, although for the reason of having helped the side opposite Zeus in the fight against the Titans. As we will see in the following episodes, that same name "Atlas" is similar to "Atla", Viking deity of the sea, direct descendant of the god of the sea and giant, Aegir. Thus we also find Atlanteotl, from the culture of the Mexicas, which is the region "surrounded by water" from which the Aztecs came (from Atzlan, Atlan, Atlante, Atlanteotl or Atlántida). Now, continuing with the concept of the beast, this aquiferous being mentioned in long legends and that dwells in "*the depths of the abysses of the seas*" (Genesis 1:21; Job 3:8; 41:1, Psalm 74: 14, 104:26 and 148:7), was called Tiamat in ancient Sumer, and was also associated with Ki or Ninhursag, always described as a giant reptilian goddess (called Cipactli by the Aztecs and Tlaltecuhtli by the Mayans). Said gigantic underground monster has parallels with a Norse myth, about the Midgard Serpent, called Jörmungandr, who is said to die in the final battle, in the "valley of the world", against Thor – both will perish in combat. This serpent is not the only one, but it is specially fed and fattened for that day: «... *and the angel of peace who was with me said to me: 'These two monsters have been prepared for the great day of God and are fed so that the punishment of the Lord of the spirits will not fall in vain on them, they will make the children die with their mothers and the children with their fathers and then the judgment will take place according to his mercy and his patience.*" (1 Enoch 60:24-25) Another prophet also spoke about this devouring serpent from the underworld, saying: «*[I proceeded] with the angel from that place about 185 days' journey. And he showed me a level ground, and a serpent, which seems to be 200 plethra in length. And he showed me Hades, and its appearance was dark and abominable. And I asked him, "who is this dragon, and what is this monster that surrounds him?" And the angel said: "The dragon is the [eternal one, who for] years eats the bodies of those who spend their lives [in] wickedness, and he feeds on them. And this is Hell, which in turn also resembles it ..."* (3rd Baruch 2-9) Despite seeming to speak of the same creature, it is very possible that many enormous forms were referred to since ancient times, but always represented as monsters or dragons.

Baruch, on his journey, asked about the beast of Hades: «*I Baruch asked the angel: "allow me to ask you one thing, sir. About what you told me about the dragon, [whose] drinks [are] a cubit from the sea, [can you] also tell me, what is his big belly like?" And the angel said to me: "His belly is Hades, and as far as a fall of 300 men is thrown, so big is his belly".*» (3rd Baruch 5:1-3) That beast, in the case of Baruch, is referred to as the immensity of the Abyss itself and linked to death as the devourer of souls. Could there be a relationship between these matters and the son of Satan, referred to in the book of Enoch: <<*[the] spirits and evil demons, and [the] blows to the stomach when it falls and blows to the soul, the bite of the Serpent, and the blows of Midday, [and taught that] [the] son of the Serpent is called Tabaat.*" (1 Enoch 69:12) The prophet Daniel also seems to have prophesied about various beasts, but John pointed out about the most terrible: « *I stood on the sand of the sea, and saw a beast rise from the sea, having seven heads and ten horns; and on his horns ten diadems; and on their heads, a blasphemous name.* » (Revelation 13:1) These horns are reminiscent of the 10 regions into which the US wants to be divided in a few years – instead of 50 states -, and where a "governor" would be placed for each region (although not does not have to directly mean this, but rather be an association). This Beast was already mentioned by the prophet Isaiah when he was visualized over the forces of the sea ruling in this world: «*And he called him in the presence of Isaiah, son of Amos, the prophet, and in the presence of Josab the son of Isaiah, in order to offer the words of justice that the king had seen: And of the sentences and the eternal torments of Gehenna, and of the prince of this world, and of his angels, and his authorities and his powers.*» (Ascension from Isaiah 1:2-3) Sakla (Samael) and Belial ("Belias" or "Beliar") seemed to continually work together, the two inseparable dominating the Abyss. The prophet Isaiah is the one who confirms that Belial, one of those from Sakla, is the "prince of this world", when he said: "*And Manasseh turned his heart aside to serve Beliar, because [he is] the angel of lawlessness, who is the prince of this world, is Beliar, whose name is Mantanbukus.*» (Ascension from Isaiah 2:4) And another example was made by Paul when he compared two antagonistic leaders: «*And what agreement does Christ have with Belial? Or what part has a believer with an unbeliever?*» (2 Corinthians 6:15) If he is a prince, it means that he has a king over him, and therefore he may also be Beelzebub (Beelzebub or Baal-Zvuv), the "prince of demons", or that Satan is linked with him.

Later, anticipating the murder of the prophet Isaiah, it was written: " *But Beliar was filled with great anger against Isaiah because of the vision [which he had seen], and because of the exposure with which he had exposed [to] Sammael, and because through him the departure of the Beloved from the seventh heaven had been made known, and of his transformation and his descent and the likeness into which he was to be transformed [which is] the likeness of man, and the persecution with which they must to be persecuted, and the executioners, [by means of which] the children of Israel will torture him, and the arrival of his twelve disciples, and of his teaching, and that it must be before the Sabbath crucified on the tree, and must be crucified along with evil men, and who must be buried in the grave [...] and risen again.*" (Ascension of Isaiah 3:13) In addition to exposing that Isaiah knew in detail what would come to the Messiah and exposed Sakla, here it is again implied that Sakla/Samael and Belial are two different characters working together. But Isaiah went further, announcing that Belial would come to manifest as a human and take his place in the world government of Earth, as an artist of deceitful miracles to support his father, the Beast: «After [time] is finished, *Beliar, the great prince, the king of this world, will descend, who has ruled the country since it came into being, yes, he will descend from his sprawl in the figure of a man, a lawless king, his mother's murderer: she herself [was] the one who [put] this king .*» (Ascension from Isaiah 4:2) Although the "accuser" was described by the prophet Zephaniah in a trip he had to Hades, thus: «At that very moment I *got up and saw a great angel before me. His hair spreads out like a lion's. His teeth were out of his mouth like a bear. His hair spreads like the woman. His body was like the snake...*" (Apocalypse of Zephaniah 6:8) And when asking the angel Eremiel about the identity of that man, he answers that *"this is the one who accuses men in the presence of the Lord."* And he adds: «*... you have prevailed and you have triumphed over the accuser, and you have reached Hades and the Abyss.*» Accuser in Hebrew is called "Ha-Shatan", which is why there is a close link between Satan and Belial and Sakla.

How did evil start if the Eternal Father is perfect and good? «*The great angel, intelligence, answered me: "In the bosom of the infinite aeons in which incorruptibility is found, wisdom, the so-called Pistis, wanted to produce a work by herself, without her spouse. His handiwork turned out as a [mere] likeness of heaven. [It is to be known that] there is a veil between the higher realities and the lower aeons, and that a shadow came into existence below the veil, and this shadow became matter, and this shadow was thrown to a particular place. Well then, the creation [of wisdom] was a work carried out on matter, a kind of abortion. Received figure from the shadow. It was an arrogant lion-like beast. It was androgynous, because, as I already said, it came from matter."* (Egyptian Coptic Gospel of Rebellion 1:30-31) The story is retold over and over again. On some occasions different versions give other nuances to better understand the narration. All things are perfect in themselves when they keep order with what comes from the Father, on the other hand, when there are incorrect mixes or protocols that are not established, that is when failures and disorder occur. An animal is pure in its kind as long as it is not mixed with a dissimilar creature, even if it is chromosomally compatible, such as horses: horses, zebras, donkeys, etc. Hybrids can arise, but these can no longer reproduce. Just like a clone, it deteriorates extremely quickly, because it is not a pure creation (that has its explanation at the cellular level). The same with food: the more different elements are mixed, the sooner the food will be damaged, which is why organic waste has a rapid decomposition process. In an annexed sense, the parameter of a feminine and a masculine force follows the order of Creation and is the way in which everything acts in accordance with the Father. By not following this guideline something arises not according to the celestial canon. In other words, creating is not difficult, the problem is that everything that is not done according to the established spiritual union, in harmony with the Father, comes to lack the divine spark of love, light and purity. That is what Jesus was referring to in his debate with wise men from the East, when they asked him about the origin of evil and he replied: « *All created things have their own colours, tones and shapes; but some tones, though good and pure in themselves, produce disharmonies and discordant tones when mixed*." (Life and Works of Jesus in Tibet and West India 4:12)

How did all that insurrection happen? That inappropriate procedure must have been described at various times: «*Although everything that is divine and human in the world affirms that [nothing] exists before [...] chaos, I, on the contrary, maintain that the whole world has [made] the mistake of ignoring the [nature] of chaos and the root. [...] all human beings [err] in the fact that each of [them ignores what] darkness is, it is a shadow [...] it was called "darkness". Shadow or an existing work from the beginning. Therefore, it is clear that [what] existed before [everything] was chaos and that we only [see what we see of] the [universe] after the first work that has arrived. Penetrating, but in truth, as in the first ones that have been translated into chaos, and show the demonstration of the truth.»* (Written without title. Treatise on the origin of the world, NH II, 5. Coptic text conserved in France) The power of Wisdom was surprising, but it was delimited, giving it a place below the Superior Realities. That creation is part of what we know today, but the other segment, the bad one, is under the Earth. However, that embodiment manifested a creature in her, which accumulated in itself a large part of that power that had come out of Wisdom: *«But due to [what] Faith-Wisdom wanted, which lacks the model of the mind and he was carrying out the government over the matter and all his powers, an Arcane came out of the water, similar to a lion, androgynous, with great power, but without knowing where he was born. And when Faith-Wisdom saw him go to the bottom of the water, she said to him: "Young man, cross to date!" Whose equivalent is "yaldabaot".»* (Apocryphon of Coptic Creation 100:1-14) Another writing relates that this Yaldabaot has a certain parallelism with the character that Ezekiel prophesied (chap. 28:2-9) -although the popular notion assumes this verse to Satan-, according to which it reads: *«[This creature] opened its eyes and saw an enormous expanse of infinite matter. Then he proudly exalted and said: "I am god and there is no other besides me." By saying this he sinned against The All. Then a voice came from Above, from the Supreme Power, saying: "Erras Samael" – that is, "the god of the blind".»* (Egyptian Coptic Gospel of the Rebellion 1:32-33) Two names are already identified for this figure: Yaldabaot and Samael. Then the classic name appears: Sakla. With these three nicknames is how this character is first known.

Another text, of a different origin, and which we have already mentioned, tells us: «*Now Sophia, who is the Wisdom of the Afterthought and who represents an Eternal Kingdom, conceived a thought. She had this idea herself, and the Invisible Spirit and the Former Thought were also reflected in her. She wanted to give birth to a being like herself without the Spirit's permission (the Spirit had not given approval), without her lover, and without her consideration. His partner did not approve; she found no one who agreed with her; and she considered this question without the permission of the Spirit or any knowledge of what she had decided. However, she gave birth to a Kindred. And, due to the unconquerable power within her, her thought was not an idle thought. Rather, something came out of her that was imperfect and in appearance different from her, for she had produced it without her lover. It did not look like its Mother, and it had a different shape. When Sofia saw what her thought had produced, she turned into the figure of a serpent with the face of a lion. His eyes were like flashing lightning. He threw it away from himself, out of that realm, so that none of the Immortals would see it. For he had produced it ignorantly. He surrounded it with a shining cloud and placed a throne in the middle of the cloud, so that no one would see it except the Holy Spirit, the one they call the Mother of the living. She named her Kindred Jaldabaoz.*" (Secret Book of John 6:1-9) Apparently, the apparition that came out of that water had the resemblance of something like a lion and also a serpent. That being did not know where it came from, but it was believed that it was the only thing that existed, that it was God, and he expressed it this way when they scolded him for saying that it was the only god: «He said: "If there is another being before me, let *it reveal me*". *At once Wisdom stretched out her finger and introduced light into matter and pursued it downwards to the regions of chaos, then ascending towards its light. Again the darkness [...] in matter. This archon, because he was androgynous, produced for himself a great aeon, an infinite greatness."* (Egyptian Coptic Gospel of the Rebellion 1:34-35) So now Samael had seen something that manifested from Above, where clearly everything is pure light, and this chased him towards the regions of the Abyss.

From Above there was more activity to reduce Samael and what he created. This situation worried more than one: "*I was thinking of the produced Ennoias who came out of the Spirit without stain, on the descent into the water, that is, the regions below.*" (2nd Treaty of the Great Set) Although before pronouncing these foolish words, Samael saw the creative capacity that was in him and it did not take long to put it into operation: «"Jaldabaoz is the first ruler *who received power from his Mother. Then he left her and wandered away from the kingdoms where he was born. He was strong, and he created for himself other realms by means of a bright flame of fire that still exists. He mated with the Foolishness in him, and produced his own authorities.*" (Secret Book of John 7:1-3) The reference to the " *bright flame* " or "*eternal fire*" is seen in several accounts, and seems to correspond to a primordial force or essence that is located somewhere, which has the ability to sparkle the mind and create life. Although the existing forces have the ability to create influences, in the form of energies, which can be their own qualities, vices or traumas, in the case of Jaldabaoz-Samael, it was with his own Stupidity. There are two ways of saying "stupidity" in Hebrew: "tipshut" and "eilut". The root of "eilut" can come from "lailut" (night), such as Lilit, Lilitu or Laila. This makes a lot of sense, since there is a close relationship between Sakla and the night, to whom Hebrew legends assume the characteristic of being a demon mother of yesteryear, symbolized by the owl. She is also understood as the personification of lust, something reminiscent of the myth of Eros and that could have something to do with copulation and sexuality in the elements.

The creations, to be correct and complete, must be carried out with the other party collaborating, hence the spirits need to mate with opposite forms (masculine with feminine or feminine with masculine) to develop some creation or living form. At that moment, however, it was when the creation of beings that later the ancient peoples venerated took place: « *That day came the beginning of the words that reached the gods, angels and men. And what happened is through speech achieved by the gods, angels and men. But Yaltabaoth Arcane was ignorant of the power of Faith. He did not face living his life, but reflected the one that had spoken in the water.*» (Coptic Creation Apocryphon 100:14-22) Samael was not able to create things by his own inspiration, because he was not part of the imperishable immortals, so upon seeing Wisdom he was inspired from there to create similar forms, but that they would not stop being vain replicas of a mere image, since the spiritual essence could not plagiarize it. She allowed herself to be seen to ridicule him. The gods had appeared behind him, partly because of the power that still flowed from those waters that Wisdom had ignorantly created, but Samael, seeing his mother, tried to create angels, powers and authorities in the likeness of the figure that was reflected in the water.

Even to top it off, what Samael saw was not a direct image but the simple reflection of Wisdom in the water. She descended when he said that he was god, so when he saw his initial creations, they laughed at him, since he had said that he was the greatest and the only thing. His first idea arose within him, for it was natural essence that he had received from his mother's power: "*Jaldabaoz organized everything according to the pattern of the first eternal realms that had been born, for he wished to create beings that were like the Undying Ones. It is not that he had seen the Undying, but that the power in him, which he had taken from his Mother, produced the pattern for the order of the world.*" When he saw the creation all *around him and the multitude of angels around him who had come out of him, he said to them: "I am a jealous god, and there is no other god besides me." But by making this statement he suggested to the angels who were with him that there is another God. For if there were no other God, who would you be jealous of? »* (Secret Book of John 7:27-30) To top off Samael's foolishness, that voice that scolded him that he was wrong argued another detail: man already existed, before him: «*And after the foundation of the world], Saclas said to his an[gels: "I], I am a Jealous God[o] and apart from me no[other exists", since he believed in his reality. Then a voice came from on high saying: "There is Man and the Son of Man", because of the descent of the image from on high, which is similar to his voice on the height of the image that he has seen. Through the vision of the image from above, the first creature was shaped.*>> (Gospel of the Egyptians, from the Great Seth 1:58-59) Could the humiliation of Samael –here called Saclas- be greater? Not only did he err when he said that he was god and that he was the only one, since the immortals already existed, but that man also existed previously, and was superior to him. That is why Seth also wrote on a later occasion that, after his words of ignorance about the deity, and being warned in reference to divinity, " *he went on to say: 'Who is man?' And all the crowd of his angels, who had seen the man and his dwelling, laughed at his littleness.*" (2nd Treatise on the Great Set) It was after acting like this that Wisdom blinded all of them, allowing herself to be seen: «*She taught these things, and revealed herself in human form*». "*The entire kingdom of the first ruler trembled and the foundations of the underworld shook. The bottom of the waters above the material world was illuminated by the image that had appeared. When all the authorities and the first ruler looked at this apparition, they saw the whole background because it was illuminated. And through the light they saw the shape of the image in the water*".» (Secret Book of John 8:13-15)

When Wisdom observed what had happened, she realized the obvious: the light had diminished: «*"Then the Mother began to move from one side to the other. She realized she was missing something when the brightness of the light dimmed. She became obscure because her lover had not collaborated with her."* (Secret Book of John 8:1) That definition of "moving" from one side to the other is what Moses referred to, who had not been given the details regarding said events: «...Ruach Elohim merajefet al *pnei ha maim*" (Barashit A:B) This from Hebrew translates that after the chaos and desolation " *the wind-spirit of the gods fluttered over the face of the waters* " (Genesis 1:2). Ergo, Jesus intended to explain to his disciple John: «*I said: "Lord, what does it mean that she moved from one side to the other?" The Lord laughed and said: "Do not suppose that it happened just as Moses said, <<on the waters>>. No, when she recognized the evil that had taken place and the theft that her son had committed, she repented. Although in the darkness he had forgotten his ignorance, he began to feel ashamed and agitated. This agitation is the movement from one side to the other." "The Arrogant took power from his Mother. He was ignorant, for he believed that no other power existed except his Mother. He saw the host of angels that he had created and exalted himself above them.* » (Secret Book of John 8:2-5) Although, one of the several texts of the Nag Hammadi Library, says that « the *archon planned to create children for himself, and he created 7 sons, who were androgynous like their father. And he said to his sons: "I am the God of all." Then Zoé, the daughter of Pistis Sofia, cried out and said to him: "You were wrong, Saclas" – whose interpretation is "Yaldabaot". Then he blew into his face and his breath turned into an angel of fire. And this angel bound Yaldabaot and threw him into Tartarus, to the place that is under the abyss.* » (Egyptian Coptic Gospel of the Rebellion 1:35) What will then have happened to him there when he was thrown by the angel? It is evident that although some stories vary in relation to others, the story is the same, since only the perspectives, the vocabulary, the perception according to the time and the culture, and the forms, often allegorical, metaphorical, literal or symbolic, change.

About Sofia, Sakla and this event, Sem was told: «*She became strong because of the light of the Spirit that was in nature. His image appeared in the water in the form of a frightening animal with many faces, which is then distorted. A light came down into a chaos full of mist and dust, in order to do harm to nature. And the light of surprise that is in the central region came to him after he got rid of the burden of Darkness.*" (Paraphrase of Sem 15:10) The point is that, although Sophia is the one who is known from the stories of the New Testament as the Holy Spirit, what is represented here warns that the innate Mind of the world in appearance had a pure creative power since it was part of the spiritual force of light from Above, being understood by certain ancient Egyptian philosophers as "the offspring of the Most High God" -since they considered the Earth as a living body and a deity, exponent of the human figure and prior to the beginning of material man. However, different from how Christianity represents it, the Holy Spirit would not be masculine but feminine, and would personify wisdom and faith, motor and coordinator of the issues to be solved, being consistent with what was manifested. Felipe wrote some lyrics in consonance with Sophia and how she deceived the gods of Sakla, saying: «*The Archons believed that by their strength and by their will they did what they did; but it is the Holy Spirit who operated in everything secretly through them according to his will. They sow everywhere the truth, which exists from the beginning, and many contemplate it as it is sown; but few who behold it reap it. Some say that Mary has conceived by the work of the Holy Spirit: they are wrong, they do not know what they are saying. When has a woman ever conceived of a woman? Mary is the virgin whom no Power has stained. She is a great anathema for the Jews, who are the apostles and the apostolics. This virgin that no Power has violated, [...while] the Powers became contaminated."* (Philip 1:16-17) Then he said: «... *The Holy Spirit feeds everyone and exercises his dominion over [all] Powers, the same over the docile ones as over the [unruly] and solitary ones, for he [...] confines them so that [...] whenever they want.»* (verse 40). With this he makes clear the role of Sofia (Wisdom and the personification of the Holy Spirit) deceiving the gods or arcana, while they believed that they were doing their will from the beginning.

Therefore, Jesus revealed to his brother James about Sakla and his archons that «*they intend to cause disturbance and violence [...] but [...] no. [...By] which you were sent with the order to produce this current [creation]. But afterwards, [when he] is put to shame, he will be puzzled, [by the fact] that his effort is far [of] aeons, it is a nothing. And his inheritance will turn out to be small, that is, the one he boasts about, [saying] it was great. His gifts are also not beneficial. Your promises are bad advice. [...] Rather, he exercises violence against you. He intends to do us injustice. And he is going to reign for some time, fixed for him.»* (2nd Apocalypse of James 52-53) Before, resurrected, Jesus had already warned his apostles: «*As I have already said, from the aeons of water on the radiations of light, a drop slipped, a drop of light and of the spirit it flowed to the lower regions of the omnipotence of chaos, so that the forms modeled from that drop can be seen, for this constitutes an act of accusation against him, the first genitor, the one who is called Jaldabaot. And this drop has revealed the shapes modeled by the spirit, giving them a living soul with its breath. For it had grown cold and fallen into the ignorance of the soul. Animated by the breath of the great light of the male, and when (Adam) began to think, that immortal being, when the breath breathed into him, named all those who are in the world of chaos and all the things that are in him. find.»* (Sheneset Gnostic 1:20) However, despite this mistake in Wisdom, she repented and no one from the Eternal Realms criticized her or rebuked her.

Rather, they helped him to restore the deficiency that had occurred, since there was a plan so that from that moment on, matter, the race of men, would manifest itself to the new plane. Ergo, I want to emphasize this position of the immortals, not going against the Mother, Wisdom, showing that they are not men nor does injustice or hint of evil dwell in them: «"When the Mother realized that this dark shadow had been born *imperfectly , she understood that her lover had not collaborated with her. He repented with many tears. The whole realm of the invisible virgin Spirit Fullness heard her prayer of repentance and offered praise for her, and the Holy Spirit poured out part of his Fullness upon her. Well, her lover had not come to her before, but now he did come to her, crossing the kingdom of Plenitude, in order to be able to restore what she lacked. She ascended not to her own eternal realm, but, instead, to a position just above her son. She would remain in this 9th heaven until she restored what was lacking."* (Secret Book of John 8:6-9)

Let's see some highlights mentioned: we can see that what Adam "named" of the "fish" and "birds" -supposed- were, in fact, the living beings according to their qualities. It is also read that the Holy Spirit is the one who pours "part of his fullness" on Sofia, which would explain why Sofia was also the Holy Spirit, the revealed one, while the original and from Above is still in the rooms of the Imperishable Universe. Hence, there is Sofia the great and Sofia the little one. It is also made clear that Sofia created thinking that, although she did not ask for approval, she would be supported as something natural and her own in her Kingdom, but it was not like that. Now Sofia had to restore the problem she had generated, since that was where the great evils that Sakla and her authorities "brought forth" later came from, among which was later the provocation to sin, that is, harming oneself - a plan perpetrated to to harm man-: «*Peter said to him: "Since you have explained everything to us, explain to us this also: what is the sin of the world?" The Savior said: "There is no sin, yet you commit sin when you practice the works of the adulterous nature called 'sin'. For this reason, goodness came among you, towards what is proper to all nature, to restore it to its roots". He went on and said: "For this reason you get sick and die, since [you practice what leads you astray. Let whoever can understand] understand. [Matter engendered] a passion lacking in similarity, since it proceeded from an act against nature. Then there is a disorder in the whole body. For this reason I told you: Be in harmony [with nature], and if you are not in harmony, you are in harmony with the various similarities of nature. Whoever has ears to listen, let him listen."* (Mary Magdalene 1:7-8)

The Diabolical Triad

The trinitarian concept is quite remote, but it appears embodied in polytheism with ideas that Rome later absorbed, incorporating them into Catholicism. In the satanic system, this point is clearly reflected in consonance with the three apocalyptic figures that will play a transcendental roll at the end of time. If it were not for John, son of Zebedee, who received the magnificent revelation at the end of this century, we would not have many very important references to describe the organization of evil. From the diabolical triad we then have the Dragon (Leviathan, the Ancient Serpent), the Beast (Behemoth) and the False Prophet, defined as "the prince of this world" –although this appellation is sometimes also related to the Devil. They all represent Satan, Sakla, the Beast, the Dragon, etc. All are one and the same, as Christ, the Father, the Holy Spirit, the angels of the One and those of Christ are one. The False Prophet precisely has *"two horns"* (Revelation 13:11) because possibly one represents the system and the other represents himself, Belial himself. If my thesis is correct, Sakla would be the Beast that will rise from the Abyss, its promoter will be precisely the latter's youngest son, Belial, known as the False Prophet, and then Satan, who is the Dragon, would come down from heaven, to be all three in its full apogee in those 3 and a half years - as we have detailed in "Armageddon E-5" and in "Recognizing the Time of the End". My perspective on the matter is that before Christ, many things about the universe that is above "the veil" were not revealed, much less the hidden mysteries in relation to the intricate organization chart of the satanic system. The reason may be quite obvious, making it clear that Christ exposed the principalities and powers and manifested the light of truth, making known what was hidden or could not be revealed until the death of the Son on the cross and his Resurrection, thus breaking the molds of power, force, law and parameters controlled by the powers and their subordinate spirits.

Regarding the fact that the villain is accompanied by the "angel of fury" -possibly the Nebruel of the Apocalypse of Set-, there is also a writing by a former member of the Sanhedrin, and a disciple of Jesus, who apparently wrote: «*And, while all the ancient fathers rejoiced, behold, Satan, prince and chief of death, said to the Fury: prepare to receive Jesus, who boasts of being the Christ and the Son of God, and who is a man of great fear of death, since I myself have heard it said: My soul is sad to death. And then I understood that he was afraid of the cross. And he added: Brother, let us get ready, both you and I, for the bad day. Let us fortify this place, to be able to hold here prisoner the one called Jesus who, according to what John and the prophets said, must come to expel us from here.*" (Nicodemus 21:1-2) It is clear that this villain did not seem to know what he was saying about Jesus, this is what they supposedly heard him say, and the resurrected witnesses who affirmed this before the Sanhedrin, would have also exposed: «And, *while Satan and the Fury thus spoke, a voice like thunder was heard, which said: Open your doors, you princes. Open, eternal doors, that the King of Glory wants to enter. And the Fury, hearing the voice, said to Satan: Go, go out, and fight against him. And Satan came out. Then the Fury said to his demons: Shut the great bronze doors, shut the great iron bolts, lock the great locks, and all of you stand sentinel, for if this man enters, we are all doomed. And, hearing these great voices, the ancient saints exclaimed: Devouring and insatiable Fury, open the King of Glory, the son of David, the one prophesied by Moses and Isaiah. And again the voice of thunder was heard saying: Open your eternal doors, for the King of Glory wants to enter. And the Fury cried out, enraged: Who is the King of Glory? And the angels of God answered: The mighty and victorious Lord. And immediately the great bronze doors were blown to pieces, and those chained by death rose up. And the King of Glory entered the figure of a man, and all the caves of Fury were illuminated. And he broke the ties, which until then had not been broken, and the help of an invincible virtue visited us, who were sitting in the depths of the darkness of our faults and in the shadow of the death of our sins."* (Nicodemus 22:1-9) Although this story seems to have quite notorious reminiscences of pomp, it can have very accurate aspects in relation to what happened at that time.

This demonic triad is the one that has to disappear definitively when the prophesied end of times comes: «*Then the great war will come to man that will make the Earth tremble and tremble in its course. Yes, then the Dark Brothers will open the war between the Light and the night. When man again conquers the ocean and flies in the air with wings like birds; when he has learned to harness enlightenment, then the time of war will begin. Great will be the battle between the forces, great the war of darkness and Light. One nation will rise up against another nation using the dark forces to destroy the Earth. The weapons of force will annihilate the man of the Earth until half of the races of men have disappeared. Then shall the Children of the Morning arise and give their decree to the sons of men, saying: O men, cease the fight against your brother. Only then can they reach the Light. Leave your disbelief, oh my brother, and follow the path and know that you are well. Then men will stop fighting, brother against brother and father against son. Then the ancient home of my people will rise from its place beneath the waves of the dark ocean. Then The Age of Light will be developed with all men seeking the Light of their goal. Then the Brothers of Light will rule the people. Banished will be the darkness of the night."* (Table 12. Emerald Tablets of Thoth) It is significant that 3 important characters are mentioned in the concluding events: the classic Satan, then Belial -who is referred to in Scripture several times as the "prince of this world"- *and* the third, the one who " *was and is not* ", and who seems to have lost the great power he received from his mother. That being, Sakla, after losing power, which passed to Nature first and then to human DNA, ended up being a prisoner of his own pride, apparently, being bound by the Earth's own elements, which would explain why the Illuminati constantly try to open dimensional portals on Earth and take out demonic beings from other planes: they try to take out pan-dimensional or trans-dimensional beings and free Sakla, as if he were held back by symbols, forces, seals, laws and energies of this planet. That explains why the globe is being affected, especially causing earthquakes. That could be referred to in the verse that points to the dark one, which is: " *He who took for himself the power of Darkness that [is] locked in the midst of his members."* (Paraphrase of Shem 11:1-5) Since the Earth is a body, and thus full of organs and members, this may be a reference to being bound in Nature, in its entrails.

The change of things from the coming of Jesus Christ is remarkable. While some issues were revealed from Abraham and others from Moses, with Jesus the mysteries began to be revealed in a significant way. Until Christ, only a few privileged people at certain times experienced visions of aspects of the future, specifically the Israelite prophets – although information was also given to other peoples. Precisely some 60 years after Jesu s' assumption is that his disciple John receives the vision of the end times with vast references never before announced. Only Enoch, Daniel, Baruch, and a few other fellow Israelites were able to catch glimpses of mysterious things. Juan had knowledge about aspects not emphasized before, not even mentioned to his people. It was revealed to the believers, since then, that the forces of evil had many ramifications and modes of action, a fact that had been lost for some time. This son of Zebedee learned that "the evil one" is a group of satanic leaders, one of them would be the Devil, but others also appear on the scene, essentially at the end of the designated period. John is told about the reign of 3 important characters after the Rapture of Christ's chosen ones: Satan, the Beast and the False Prophet. Certainly we have already mentioned this in eschatological books, but I will refer to specifying how the Serpent, *"which is the devil and Satan"* swarms through space and descends to Earth when it fails to stop the Rapture. He is pointed out as *"the ancient serpent"*, in association with the sin that he provoked in the Garden of Eden and other aspects of that time. He seems to lead a team of 7 characters, exactly like the Beast. It is not clear if those 7 of Satan are the same 7 of the Beast or if they share an interest in this sacred number – later used in the occult. The same can be seen with the 10 horns and the 10 diadems, the same pattern that both obey. So, while Satan is thrown out of heaven by force, thanks to the intervention of the archangel Michael, the Beast *"comes up from the abyss"* (Revelation 11:7), but why does he hardly come up at that time? Why is it not spoken of in previous centuries? Why hadn't he gone up before? Some would say that he was waiting for the right moment, others would suggest that he is released from his place of confinement. In a later description, John would observe that Beast " *rising out of the sea* " (Rev. 13:1), understanding that the sea continues to be a clear identification of the "depths", as we have previously mentioned. In the time of the prophets, some like Daniel, Ezra and Baruch gave similar descriptions of this "*fourth beast*" (Daniel 7), very different from the previous ones and, apparently, contemporary with them.

In these accounts the writer of the Apocalypse does not ask many questions, possibly because of his surprise or because he had already been informed of aspects that he was observing. For this reason it is not clear how it is that, on the Beast, "*one of its heads*" had been "*likely wounded to death*"; Looking for the answer could be found in the legendary battles of Jehovah, when he divided Leviathan (Dragon) from Behemoth (the Beast), according to my theory. This is noticeable, for example, in Psalm 74:13-14 (in the Greek and Latin versions it is at 73:13). This is how the beast that John saw «*was, and is not; and he is about to rise from the abyss and go to perdition.*" (Apoc. 17:8) It is important to note that at present, and for decades now, those dark ones that are at the service of evil open dimensional portals in specific places on Earth. This entire planet is one body, a web of life interconnected by channels not visible to the human eye. These connections follow toroid-shaped patterns and intertwine with all the cardinal points and stellar alignments. It is as if the entire universe were a virtual machine or gear under exact numerical standards, following precisions duly established before existence and still fitted with seconds, minutes, hours, days, months and years, exactly. On Earth these points are defined as Ley Lines, and it is on them that the great monuments, pyramids, stadiums, churches, monasteries, parliaments, skyscrapers, abbeys, mosques, etc. have been erected. These sites seem to have supernatural energy vortexes that lead multiple sectarian groups to gather there to perform all kinds of rituals. Precisely the appearances of the Crop Circles (Crop Lines) in the United Kingdom obey points of the Ley Lines. These locations are used to try to remove all kinds of dark beings from the bowels of the Earth, which would explain the departure of the Beast of his place. A similar example already occurred around 10,500 BC, in Atlantis, but that will be another story that we will deal with in the next volumes. Ergo, pondering, we analyze that this Beast corresponds to Sakla, the father of the 7 and part of them. Hence John was told: « *The beast that was, and is not, is also the eighth; and it is among the seven, and it goes to perdition.* » (Apoc. 17:11) For their part, the 7 are not spoken of as specific characters, as if, possibly, they were oneself with their representative or simply were not there anymore, only the vision of these beings being an identification of that is. That responds to why Belial fits with the figure of the False Prophet (the Islamic Dajjal), being that he is one of the original 12 of Sakla, which were divided into two groups: those of the heavens and those of the underworld. Where are those 7? Enoch mentions 7 imprisoned for thousands of years, but who knows if they are the same (see 1 Enoch 18:12-16). And where are the other 4 that are not mentioned? Enoch speaks of 4, precisely, locked up for 10,000 years (see 1 Enoch 21:1-10). Belial would be one of the 5 sent to the underworld and the lord of it, reserved only a little more for the antichrist apogee.

Although starting from the year 2013 there are still a few years left for this culmination to take place, the end of these individuals is well known thanks to the Apocalypse of John, inaugurating hell, that is, Gehenna: «And the beast was captured, and with *it the false prophet who had performed before her the signs with which he had deceived those who had received the mark of the beast, and had worshiped his image. These two were thrown alive into a lake of fire that burns with brimstone.*" (Revelation 19:20) If not referring to Sakla's former situation, this verse may be speaking of its final annulment: "*Before the end of [times], the whole place will be struck by lightning. Then the Arcane will mourn, [his crying] [and] death. Angels will cry over their men and demons will cry over their time and men will cry and cry over their death.*" (Apocryphon of Coptic Creation 125:32 to 126:4) The death of Satan will also come ten centuries later: « *And the devil who deceived them was thrown into the lake of fire and brimstone, where the beast and the false prophet were; and they will be tormented day and night forever and ever."* (Revelation 20:10) All this corresponds to the repeated words that remind us how this whole story will conclude: « *Then begins the age of years to come and that will be thrown into chaos. Their kings will be drunk on the flaming sword and there will be war between them so that the earth is drunk with paid blood and the seas will be torn apart by combat. Then the sun will be darkened and the moon will lose its brightness, the stars in the sky turn away from their course.»* (Coptic Creation Apocryphon 126:4-12) And this story ends like this: «*And there will be a great thunder, of a great power that is above all the powers of chaos, where the firmament of the woman is. Having created the first film, he vows to bring in wise intelligence [of] fire and take insane anger. Then she will follow [the] chaos gods she has created, and the Great Bull. From it they sink into the abyss. They will be wiped out because of their injustice, they will be like these mountains on fire and [will] eat each other until they are destroyed by their great parents. When they destroy themselves, they will turn [be] against [them] themselves and destroy until they perish. And the heavens will fall on each other and burn their powers. Their eons will also be troublesome. And the sky will collapse and split in two. Sound [of whose] fall [will be heard] on Earth [that] it may support them. They will fall into the abyss, and the abyss will be reversed. Light and darkness will disappear. It will be like it never existed, and the work [of] the dark will dissolve. Then the deficiency will be removed at the root, the bottom in the dark, and the light will come out [to] the top at its root. And the glory of the unbegotten will appear and fill the aeons ...*" (Coptic Creation Apocryphon 126:13 to 127:7)

Sakla created many forms through her power, but was drawn by Sophia into the Abyss and chaos (the chaotic aspect was sucked into the lower parts of this globe), created by herself, along with the rest of her emerging powers and rulers. –according to one of these Egyptian texts. Kinds of those went to stop there and Sakla was removed from heaven where he flaunted his boldness: «*When the heavens were established with their powers and the fullness of their hotel, the Great bull was glorified by the army of [his] angels. And all the gods and angels praised and blessed him. And he is [the one who] henceforth [thought] in his heart and boasted, telling them: 'I don't need anything.' He said: 'I am god and there are no others besides me.' By saying [this] however, he sinned against all the preaching immortals he saw. But when Faith saw the great impiety [of the] Arcanum, she got angry, came and said: 'You are wrong, Samael' - that is, 'the blind god' -, 'there is before you an immortal man, a man of light, which will be [put] on your model. You [will be] trampled like many potters [of] clay, and they will make you plummet with your mother into the Abyss. In fact, when your work has come to an end the entire disability will end, it will be [so] since the truth appears and [then] you will disappear, and you will become what has never been.' Having said this, he revealed the Faith in his water [the] reflection of his greatness. And she was retired and [instated] at the top of her light.*" (Apocryphon of Coptic Creation 103:3-32) The prophet Ezekiel (ch. 28:2-10) would seem to have also spoken of it, but, despite the fact that there is a certain parallelism between Sakla and the Opposer, translated as Tire, it is only clear in terms of saying that it was god, since for the rest of the appreciations, very possibly it was referring -I think- to Satan.

Let us clarify that the name usually translated as "Tyre", a region of Phoenicia, near Sidon, is sometimes written "Tzer" and other times "'Tzor", which means "Opponent" (usually translated as: Zur, South, Tire, Flint, Rock or Shelter). This character does not seem to be mentioned here with as much creative power as the descriptions of Sakla. Some believe that there are two people behind this evil figure – or more. Even the prophet continues to affirm a word against the sovereign of "Tyre", although the name differs from the previous one; It is not "Tzer" but "Tzor". This theory would be supported by the fact that on the one hand it appears that an Evil One was born by Sofia and was ignorant of everything from the beginning; He was also powerful and created great things in the universe, such as gods and angels. This theme is not supported by biblical verses but only by themes taken from Coptic scrolls, that is, Egyptian Christians, basically from those found in Nag Hammadi. Since it is understood in the other version, the popular one, that the Evil One was an angel with the rank of Cherub (from where the plural word: "cherub" comes from), who was good in everything, received homage at his graduation, being a great example to all, until evil was found in him and he deserted while in the Garden of Eden. It could be assumed that both were like the same person, as when speaking of the Creator Father, the figure of God, the Holy Spirit of God, the Elohim or Yehovah (it is believed that all these are the same being, when everything leads to the assumption that it is an assembly subject to the Great Invisible Spirit or Creator Father). This assumption would be supported by the verse we quoted earlier about Sakla and Nebruel becoming as one. In other words, according to another verse: *"he allied himself with the Stupidity that is in him "*, with which two forces or spirits were linked and working together as if they were one, but they did not necessarily have to be the only ones (they could have more than one "rebellion" took place, in different times and places, and by different creatures). Although it could be a primary quality that is with Sakla and, besides, Satan has also been there since the establishment of the Garden of Eden, the general idea always says that Satan is the bad guy, possibly because references to evil always fall on him, understanding that acts, while de Sakla seems to have been "neutralized" the matter.

There is a similar position that was proposed by the ufologist Sixto Paz Wells, who says that Satan was an extraterrestrial originating from the Orion Constellation, healthy and fair in everything, until he was influenced by Lucifer, a being from a higher dimension, the first rebel who, according to said Peruvian researcher, would be an entity understood as Hillel, or of this typology. Although, one of these figures, which represents darkness, spiritually speaking, is also related to the "Melej Babel" or "King of Babylon", which is clearly not accurate as to whether it is the same "Morning Star", when Ezekiel speaks of his death: «... *you will pronounce this proverb against the king of Babylon, and you will say: How the oppressor stopped, how the city greedy for gold ended! Jehovah has broken the staff of the wicked, the scepter of the lords; the one who hurt the peoples with fury, with a permanent sore, the one who ruled over the nations in anger, and persecuted them with cruelty. [...] and the cedars of Lebanon, saying: Since you perished, no cutter has come up against us. Sheol below is terrified of you; he awoke the dead who came out to receive you at your coming, he made all the princes of the earth rise from their seats, all the kings of the nations. They will all cry out and say to you: Did you also become weak like us, and become like us? Your pride descended to Sheol, and the sound of your harps; worms will be your bed, and worms will cover you. How you fell from heaven, oh Lucero, son of the dawn! You were cut to the ground, you who weakened the nations. You who said in your heart: I will go up to heaven; on high, next to the stars of God, I will raise my throne, and on the mountain of testimony I will sit, on the sides of the north; I will ascend above the heights of the clouds, and I will be like the Most High. But you are knocked down to Sheol, on the sides of the abyss. Those who see you will lean towards you, they will contemplate you, saying: Is this the man who made the earth tremble, who shook kingdoms? that made the world like a desert, that devastated its cities, that jail never opened its prisoners? All the kings of the nations, all of them lie with honor, each one in his dwelling; but you are thrown out of your grave as an abominable offspring, as clothed with those killed by the sword, who descended to the bottom of the grave; like a dead body trampled You will not be numbered with them in the grave; because you destroyed your land, you killed your people. The offspring of the wicked will not be named forever."* (Isaiah 14:4-20) It is curious that, if it were the same man, they call him prince and also king. However, these indicated characteristics again seem to refer to Satan, and not to Sakla, as I have already said.

Potentates and Authorities

Well, having already taken a tour of concepts such as the Father and his Kingdom, the imperishable, the heavens and the invisible creation, and also delved into who is Wisdom (Sophia) and how man existed, let's see what Sakla (Saclas, Jaldabaot, Jaldabaoz, Samael...) created after him: «*However, the archons of all Jaldabaoth were disobedient because of the Ennoia that came to him from his sister Sophia. They made for themselves a union with those who were with them in a mixture of a cloud of fire, which was their envy, and the rest who presented themselves, by their creatures, as if they had struck the noble pleasure of the Assembly.* " (2nd Treaty of the Great Set) With Sakla also arose gods, who, for their part, created other forms as their consorts, after which their assistant angels and the first demons were created. Shem was taught about "*the great powers that were in existence in the beginning, before I came forward. There was light and darkness and there was no Spirit between them."* (Paraphrase of Sem 1:20) This denotes that these beings were not from pre-existence, therefore their essence or spirit did not specifically come from Above, from superior realities, and also suggests that their manifestation took place at the same time as they appeared. the visible, our current reality. This son of Noah added: «*And darkness was the wind on the waters. His mind was wrapped in a chaotic fire. And the Spirit between them was a soft and humble light."* (Paraphrase of Sem 2:1-5) Here we see the typical reference that affirms, once again, that the motivation and reality in the waters, at that time, was, in fact, the darkness itself, that is, what had animation in the waters was something "black". What reigned and directed that question was chaos, being the only thing with divine life and celestial sense, acting there, the Ruaj of divinity, personified in Sofia, which was seen as a beautiful light. « *They each reigned in themselves, alone. And they covered each other, each with their power. However, the light, since it possessed great power, knew that the humiliation of Darkness and its disorder, that is, that the root was not straight.* » (Paraphrase of Sem 2:5-10) These helped each other, complementing each other in their own power and trying to reign among themselves, since they did not receive anything from Above or kingdom over which to dominate. Sofia, being one of the divine beings, knew that what was reflected there was not healthy or correct. This text assumes that Sakla is the personification of Darkness, the appropriate synonym for Darkness.

In any case, the first creations that came from Sakla were intended to imitate the things of the Undying Realm, so all the things he made were either limited replicas or imitations of the real thing. That first group of creatures that Sakla created, with Nebruel (the stupidity that is in him), were his androgynous children (7 potentates or powers) and his authorities (12 kings, gods or lords), which at the same time are his powers. With his 7 children, he created 7 faculties, so that in this way all these would always unite their qualities and bring other creative spirits into existence. In his case, the 7 sons of Sakla and the 7 faculties created 49 demons, and with these demons they continued to create until they formed a court of 365 angels. Some versions define its 12 authorities as Assistant Angels and as Generating Spirits of the Earth, that is, they are Sakla's helpers and they are also forces that would have generated, provoked, proliferated and motivated life in our world.

In the book kept in Coptic about the Rebellion, they talk about the names of the sons of Sakla, but they refer to some that are not among the 12 nor among the 7. The first were: "Iao" (I), "Eloai" (E) and Astafaios (How). About them, the text says more, although very poorly translated. The fact is that it is emphasized that its aspects are more detailed in the First Book of Norea, one of the daughters of Adam and Eve – a manuscript that I have not been able to obtain. It is said that Sakla created these children so that they would reign with him and dominate over the gods and angels and over all created things in the 6th Heaven. If you have always wondered the origin of the pantheon of 12 mythical beings, here is your start. The fact that there were 12 Olympians with Zeus or 12 deities of the Sumerian god Anu, has its birth with the Sakla pantheon that arose after these three beings. This system is repeated in many cultures when talking about their pantheons.

Paul revealed a great truth when he wrote: "*For we do not wrestle against flesh and blood, but against principalities, against powers, against the rulers of the darkness of this age, against spiritual wickedness in high places.*" (Ephesians 6:12) About this " *power of darkness* " (Luke 22:53 and Colossians 1:13) Jesus and his disciples spoke well. Now, the Lord, Jesus, named John the 12 rulers on which Seth, the son of Adam, also emphasized, and who are the representative faces of Sakla (who is usually defined as "Great Bull", also as " the First Ruler" and also as "the Beast"), since he never shows his face personally (although the prophecies warn that he will do so in recent times). The designated dozen are:

1. Atot or Athoth, whom generations of human beings call.

2. Harmas, the jealous eye or eye of fire. Where I think the symbol of the "eye of Horus" or "eye of Sauron" possibly comes from, in Tolkien's novel.

3. Galila or Kalila-Oumbri.

4. Yobel or Yabel.

5. Adoneo or Adonaios, who is called Sabaoth.

6. Cain, whom generations of human beings call the sun.

7. Abel.

8. Aquiresin or Abrilsene.

9. Yobel or Yubel .

10. Harmupiael or Armoupiael.

11. Arciadoneo or Melcheir-Adonein.

12. Belias, who is above the depths of Hades. This is identified as the one who will come to the surface and Christians will recognize him as the False Prophet, the Antichrist, the Son of Perdition or the Iniquitous.

Seth wrote that *"these are] the ones who preside over Ha[des and chaos]."* (The Great Invisible Spirit. Gospel of the Egyptians) We can also conclude that Sakla «*destined 7 kings (one for each sphere of heaven) to reign over the 7 heavens, and 5 to reign over the depths of Hades. He shared his fire with them, but he did not give up any part of the power of the light that he had taken from his Mother. For it is a being of ignorant darkness. When the light mixed with the darkness, it made the darkness brighter. When the darkness mixed with the light, they obscured the light. The result was neither light nor darkness, but rather penumbra. [...] The rulers created for themselves 7 powers. The powers in turn created for themselves 6 angels each, until there were 365 angels.*» (Secret Book of John 7:8-14) Those 7 spheres of heaven are none other than the planets of our solar system: Mercury, Venus, Mars, Jupiter, Saturn, Uranus and Neptune. It is equally important to see the similarity with the zodiac, since the Sumerians associated all these facts, being the closest to the oldest gods. The word "zodiac" comes from the Greek "zodiakos kyklos" (animal circle), because the design of the groups of stars resembled animals in their apparent shape. Although, those names and apparent imaginary forms actually originated in Sumer, where the 12 constellations of the zodiac were called UL.UE (bright flock), and the Anunnaki defined them like this:

1. GU.AN.NA ("heavenly bull"), Taurus.

2. MASH.TAB.BA ("twins"), Gemini.

3. DUB ("pincers", "pincers"), the Crab or Cancer.

4. UR.GULA ("lion"), in Latin Leo.

5. AB.SIN ("her father was Sin"), the Maiden, Virgo.

6. ZI.BA.AN.NA ("heavenly destiny"), the scales or Libra.

7. GIR.TAB ("what pinches and cuts"), Scorpio.

8. PA.BIL ("defender"), the Archer, Sagittarius.

9. SUHUR.MASH ("goat-fish"), Capricorn.

10. GU ("lord of the waters"), the Water Bearer, Aquarius.

11. SIM.MAH ("fish"), Pisces.

12. KU.MAL ("country dweller"), the Ram, Aries.

We cannot deny that the similarities are considerable and that the things that the ancient peoples absorbed were, in fact, the concepts of the gods. Now, on the other hand, referring to the 7 potentates and their corresponding appearances, we have:

1. Atot or Atoth, who has the face of a sheep.

2. Eloaios, who has the face of a mule.

3. Astafaios, who has the face of a hyena.

4. Yao, who has the face of a 7-headed serpent.

5. Sabaot, who has the face of a dragon.

6. Adonaios or Adonin, who has the face of a monkey.

7. Sabataios or Shabbataios, who has the face of flaming fire.

What is a «*flaming fire face*»? Or even more, how do you figure out a «*7-headed snake face*»? It should say that it is a being that has "7 heads", not a "face" that has 7 heads, unless when talking about "face" or "face" you mean the representation or identification of something. Therefore, it could be assumed that the animal forms reflect a presentation of each of these, even in the way of what each animal represents in itself. In addition to this, we should consider the fact that the Chinese calendar, which changes every 12 years, has 12 animals that identify each year (one per lustrum), which seem to have come out of this order, just like the zodiacal houses themselves. Let's analyze: in the Chinese calendar appear: the rabbit, the tiger, the ox, the rat, the pig, the dog, the chicken and, next, the monkey, the sheep, the horse, the snake and the dragon. The parallelism is the same with the sheep, the dragon and the monkey, and if we consider the similarity between the hyena with the dog, the mule with the horse, and the normal snake with the one with 7 heads, we can easily substitute them, taking 6 of the

7 animals referred to above correspond to half of the Chinese calendar; something that would be somewhat ridiculous to attribute to coincidence, since even the animals follow a correlative pattern in line with the 6 beings referred to (the only one lacking in animal aspect is that it identifies the burning fire, but that could be replaced by a similarity animal later). Also, Jesus said: *"This explains the 7 days of the week. Jaldabaoz had many faces besides all of these, so he could show whatever face he wanted when he was among the angels. He shared his fire with them and ruled over them because of the glorious power he had from his Mother's light. That is why he called himself God and despised the kingdom from which he came. He united 7 of his powers of thought with the authorities that were with him. When he spoke, it was done."* (Secret Book of John 7:18-21) He named each of his powers, starting with the highest:

1. Goodness, with the first authority, Atot.
2. Previous Thought, with the second authority, Eloaios.
3. Divinity, with the third authority, Astafaios.
4. Lordship, with the fourth authority, Yao.
5. Kingdom, with the fifth authority, Sabaoth.
6. Jealousy, with the sixth authority, Adonin.
7. Understanding, with the seventh authority, Sabataios.

It is clearly seen that he was replicating the things that were on him, including the question that he wanted them to dominate over what already existed, including those "celestial spheres", concrete planets, being that with this, once again, the investigative journey gave us It leads to astronomy and astrology to understand many ancient references regarding the stars: *« These beings have spheres in the celestial realms. The powers were given names from the glory above, but these names could destroy the powers. For while the names given to them by their creator were powerful, the names they received from the glory above could cause destruction and loss of power. That's why they have two names. Jaldabaoz arranged everything according to the pattern of the first everborn realms, wanting to create beings that were like the Undying. Not that he had seen the Undying, but that the power in him, which he had taken from his Mother, produced the pattern for the order of the world."* (Secret Book of John 7:24-28) Let's meditate on how important it was to define a name or a statement. The Most High revealed to Enoch: «*And in the fourth interval I ordered that there should be the great luminaries in the circles-spheres of the Heavens. In the circle-sphere, the first, the highest I placed the star-planet Kruno, in the second above I placed Aphrodit, in the third Aris, in the fourth Helios, in the fifth Zoues, in the sixth Ermis, in the seventh Selene; and in the stars-moons, the small, beautiful-bright ones, those that fly, the minor ones.*» (2nd Enoch 30:3-4) It must be seen that the source took these names that are clearly of Greek origin, since Kruno is the same as Cronos, the planet Saturn (in Hebrew: "Shabetai"), then Aphrodit is Aphrodite, the planet Venus (Hebrew: "Noga"), Aris is Ares himself, the planet Mars (Hebrew: "Maedim"), then Helios is the Sun (Hebrew: "Shemesh"), Zoues is clearly Zeus, the planet Jupiter (in Hebrew: "Tzedek", which means justice), Ermis is Hermes, the planet Mercury (in Hebrew: "kojab", which also translates: star/star), and finally Selene, which is the Moon (in Hebrew: "Iareach"). These seem to be located in 7 skies, one for each, which locates them in orbits – as we understand it today.

Sakla's daring had no limits, because understanding the root of the names he gave his offspring shows his interest in giving them a "majestic" quality, considering the power of the "Name" and the ability to empower a being with qualities special and supernatural forces. Some examples, to elucidate and figure out the idea, are: "Harmas", possibly derived from what would be translated from pre-Semitic languages as "the one from the mountain", accepting that the mountain is a symbol of government. Those who define themselves as Yobel, Yabel or Yubel come from the root that means: "divine master". The same is reflected with Adoneo, Adonaios or Adonin, which means "the one of lordship". Sabaot is "armies" or "hosts." Sabataios may be a deformation of "shabataio", which refers to "he in whom rest resides". The other genetic names of Cain and Abel translate "spear" and "frivolity" or "beauty," respectively, and show that Eve did not name her firstborn by her own invention. Such Harmupiael or Armoupiael may have to do with "mountain where the Most High proclaims", or "government where the voice of the Alto resides". Likewise, Eloaio refers to the one who is "most high". And another example would be Arciadoneo or Melcheir-Adonein, absorbing the first example, possibly, reminiscences of the Greek, and passing to "older gentleman", not of age but of superiority. The other version, without a Greek tendency, would refer to practically the same thing: "the king of dominion". Those gods of chaos and their powers, faculties or authorities are the ones that were humiliated and submitted by Christ according to the laws that arose from the initial chaos. For that reason, referring to Jesus, it was written: « *But the Word that is [and] of what [is] above all, was sent so that it would only proclaim what is unknown. He said: "There is nothing hidden that is not shown and what was not known is known. And of these was sent to manifest [what] is hidden, and the everlasting authorities of impiety and chaos, since they were condemned to death."* (Coptic Creation Apocryphon 125:14-23) Thus it is added on the cross of Christ: "*So when everything [that] seemed perfect in the modeling of the Arcanum and when the truth was revealed that it has no equivalent, all the wisdom of the gods was discredited, his fate was set in trials and his power died [losing it], his dominion was overthrown and [of futility] and Providence broke his glory.*" (Chap. 125:23-32) The question of these procedures is that in due time the indicated guideline is developed to get rid of the forces of chaos until they are completely eliminated: «*It is necessary that everything return to the place from which it arose each one; by his act and his knowledge, he will reveal his nature.*" (Ch. 127:14-17)

IV.
MESSY AND EMPTY?

THE CHAOS

" *The wise man never says everything he thinks, but always thinks everything he says.*"
Aristotle (384 – 322 BC)

❚❚ *Stretches the North towards Chaos, supports the Earth on nothingness.*" (Job 26:7. Hebrew version) Chaos beyond the north? We can accept that there are other dimensions, other universes, and therefore, other heavens, but what does all this lead us to? What we are most interested in clarifying today is how it all began, and how we got here: «*Although everything that is divine and human in the world affirms that nothing existed before chaos, I, on the contrary, maintain that the whole world it has erred by ignoring the [nature] of chaos and its root. [Here is] the demonstration: it is true that there is [a] [cord between] all human beings, in the fact that [in] each of [the beginnings] it is darkness, it is a shadow that was called "darkness". Shadow or an existing work from the beginning. Therefore, it is clear that it existed before what was the chaos and that it only appeared after the first work that has arrived. The truth truly penetrated, as in the first, those that have been translated into chaos, and show the demonstration of the truth.*» (Apocryphon of Coptic Creation 94:24 to 98:11) Can then chaos or darkness translate into the manifestation of evil?: «*Then a magician stood up and said: If you answer me, your problem will be solved. We all recognize the fact that there is evil. And everything that exists must have a cause, so God, the One, did not create evil, what God created it? And Jesus said: Everything that God, the One, has made is good, just like the first cause, the seven spirits are all good, and everything that comes out of his hand is good. All created things have certain colors, tones, and shapes of their own; but some tones, though good and pure in themselves, produce disharmonies and discordant tones when mixed. And some things, being good and pure, produce discordant things when mixed, poisonous things that men call bad. That is why evil is the disharmonious mixture of colors, tones or forms of good.*" (Life and Works of Jesus in Tibet and West India 4:9-14) Chaos seems to have been the oldest concept in the creation story, apparently promoted by disorderly "proceedings" – hence it is considered, in Egyptian culture, as a primordial negative personification, called Apophis. We have seen before that Wisdom brought about a different creation, something resulting from her own thought and desire, without the approval of the rest of the Assembly or her partner or the universal Father. That, it seems, caused the primordial chaos and the appearance of the primal gods.

At the time Moses wrote that the Aretz (translated as "Earth") was in "Tohú" and "Bohú", which is usually understood as "Chaos" and "Desolation" or "Darkness". In the Apocryphon of John (Secret Book of John) Chaos is spoken of as an epithet to refer to Tehóm (Abyss) or the Abyss as a later situation of chaos first, another aspect that participates in the creationist narrative. The Apocryphon of Set (Secret Book of Set or Book of the Great Invisible Spirit) does not mention the appearance of Satan or Chaos, but considers that this existential principle (chaos) was already present during the development of life in the imperishable realms: «*After 5,000 years the great luminary Elelet said: "Let someone rule over chaos and Hades."*» These fifty centuries would have elapsed after the imperishable Adam and Set received their immortal kingdoms to inhabit. Would chaos have occurred during those 5,000 years after the beginning of the reigns of that universe? Seth wrote: « *And there appeared a cloud [whose name is] material Wisdom [... she] looked at the regions [of chaos], [...]. And [the great angel Gamaliel] said [to the great Gabri]l, the servant of [the great light], Oroiael. He [said: «An a]ngel come forth [to rule over chaos [and Hades. Then the [satisfied] cloud came] into the two monads [of which] each had a light [...throne] which she had placed [above] in the cloud. [Then Saclas, the great [angel] saw, the great demon [who is with him, Nebr]uel. And they became [together a] generating spirit of the earth.*" As we can see, Sofia –as the Greeks call her– appeared when the archangels determined how to deal or direct the existing situation of chaos. Even so, the relationship of Sofia (Material Wisdom) with these matters is strange, as the Apocryphon of John says, noting that she "fluttered over the face of the water" and it was based on her image that Sakla created life, helped by Nebruel. (Other texts define the demon of this story as "*the stupidity*" that was in Sakla, that is, the personification of stupidity).

Why then did no one reject the idea of Sakla offering to rule over Chaos? It is also worth noting that Hades, according to this source, already existed at that time, but the Father, being pure, perfect, holy and good, could not create death, chaos, darkness, evil or similar things. Therefore, the Bible, in the genesis story, begins by reporting an event that had already begun: it was not the beginning as such, but the beginning of something. Although, that Hades did not seem like an infernal concept but rather a result of the separation of chaos, which Enoch describes being taken to "a hole", which is later defined as "Tehom" (Abyss), plural of "Tohu". In the region of Gadar, located on the opposite bank of Galilee, Jesus freed a possessed boy, and the information absorbed from there is substantial: «*And Jesus asked him, saying: What is your name? And he said: Legion. Because a lot of demons have entered him. And they begged him not to send them into the abyss.*" (Luke 8:30-31) This is curious, since some versions consider that the demons were deported to the Abyss, and in this one, Set says something like that they volunteered to go there, at least as far as Sakla is concerned. –apparently, before he started his own creations. However, it seems that the idea itself that they were taken to the underworld was entirely proposed and directed by Sofía. What interest did Sofia have in Sakla becoming governor of Chaos, or just going to that place? It can be assumed that she determined this as a result of what she saw in Sakla and her kind: the bad guys must be on the plane of error, not with the celestials. Possibly Greek mythology referred to this by saying that Erebos was turned into the Infernal Shadows, having been deported to the underworld to be the personification of it. Squeezing the texts of the Egyptian Library of Nag Hammadi, we get that Sofia's thought was created automatically and immediately when she wanted it, and that creation that took place was somewhat deformed: «For those who were in the world it had been prepared by the *will of our sister Sofía - the one who is a harlot - because of her innocence, which has not been pronounced. And she did not ask for anything at all, neither from the greatness of the Assembly, nor from the Pleroma.*» (2nd Treatise on the Great Set) The translation, by defining Sofia as a "*harlot*", does not denote a pretentious, tendentious, subjective, negative or obscene aspect, but rather encompasses the quality of "mixture" of forces and disharmonious action in everything that this implies.

From there arose darkness and chaos, moved by a great power that left Sofia, but had to return to her, which is why it is the Holy Spirit who intervenes in the world. That power came to produce things due to the creative capacity and magnificence that was in it, since it was the creative power of the universal Father acting, since the thought of the Father already creates in itself, just as his thought made action or later does. his mere word or derivative of himself, as are the imperishables. From there, from that cosmic apparition, the ancient peoples say that we see today the existing universe, but that it also has planes, levels, dimensions and powers that we do not know or glimpse with the naked eye. In it the gods arose, either those of chaos or the celestial ones (not yet the fools, who were really angels and extraterrestrials posing as gods), beings prior to the events of the Fall, of which legendary peoples left chronicles embodied in stone, paper, cloth, wood, clay and many other bases, so that we would have proof of it. The Egyptians argued that the original principle arose from an Ogdoad (Assembly of 8) -as the so-called "gnostic" texts also say- which was composed of the even principles of Primal Water, Infinite Space (eternity and infinite sky). Darkness (the darkness of primordial chaos) and the Occult, or the Hidden Power. All these were born, clearly, of a Primeval Chaos. If Sakla and Neburel became « *a generating spirit of the Earth* », it means that they descended into Chaos and Hades and had access to the earth's surface to develop and foster life. After this Egyptian Ogdoad, of the Primal Water (personified in Nun and his consort Naunet), the demiurge Atum (also known as "Kematef") arose, and together, the 9 principles, formed the Heliopolitan Ennead (Heliopolitan, meaning that it was venerated and respected in the City of the Sun: "Heliopolis", currently known as Cairo). Another version says that Atum arose from the head of Ptah. In this cosmogony it was believed that these 9 gods of the Ennead would be: Nun and Naunet (who gave rise to Heryshaf), Heh/Hah and Hehet, Kek and Keket (Kauket), Tenemu and Tenemuit, and Atum, although in other perspectives they were not. Tenemu and Tenemuit appeared but Amun and Amaunet. All of these were portrayed as men and women with snake and frog heads, respectively, or simply as snake and frog altogether.

Nun, the Primordial Water, surrounded the entire world, according to the myth, and from it everything was created. Ergo, we always have the origin of things in Chaos, so we can only assume that this is how our home began to be terra-formed. Chronologically, after the Ennead, came the gods and their gatherings. Osiris's brother, Seth, was known as the "god of Chaos", basically because of his instigating and negative role, just as Loki was in Norse mythology, Eris in Greek, Shivá in Hindu (the black sheep of the family). It is even known that Set murdered Osiris, dismembering him, and later Horus, the son of Osiris with his sister Isis, made war on Set and achieved sovereignty over Egypt, under the protection of Ra, and later Amun, commanding Set to the desert under exile. But the Greeks, despite apparently being a later-Egyptian civilization, also mentioned sections of those dark beginnings. The Hellenes (the name they received from the story of Helena, although it began to be used in the time of Alexander the Great in relation to the influence that the Greeks exercised in relation to their culture) affirmed that the first thing was Xaos (Chaos) and Zoton (Darkness), although Zoton or Soton was a term that God used to describe the creation of the evil: "*And he (Adam) was continually in paradise, and the Devil understood that I wanted to create another world, because Adam was lord in the Earth, to rule and control it. The Devil is the evil spirit of the low places, as a fugitive he made Sotona from the heavens as his name was Satanail, thus he became different from the messenger-soldiers, but his nature did not change his intelligence so much as his understanding of things. upright (righteous) and sinful.*" (2nd Enoch 31:3-6) Sotona, in the text of Enoch, is sometimes strangely translated as "Diana" (name of a Roman goddess that corresponds to the Greek Artemis, divinity of the Moon and twin sister of Apollo, the God of the sun). The word "Soton" has a Greek root meaning "Darkness". The name Satanail means, as it is assumed: "the impious" (Shatanael translates: "Opponent of God") In Hebrew it is: "Ha Satan", that is, "the Adversary". It is therefore interesting that it is clarified that he was "*the evil spirit of the nether places*", that is, of the Abyss, and also it is made known that "*as a fugitive*" he made or produced "*the Darkness of the heavens*". Therefore we understand that it is an aspect of spiritual vocabulary: he caused darkness in the Kingdom of God, he was not a creator, because he was not a deity, but he caused evil among those of God. What is not clear is his identity, but as in other accounts, he seems to be confused, associated, related or mixed with Sakla, assuming that both are under the same "spirit", like Christ and his followers, or Christ and the Father, who although as entities they are separated from each other, they are One.

The Greeks said that from the union or mixture of Chaos and "Darkness" (equivalent to the Hebrew "Tohú" and "Bohú") Nyx (Night) and Erebo (Darkness) arose, something quite accurate to represent the result or consequences by the hand of Sakla in the creation of God. Nyx, or "The Night", is also known by the freemasons, Hebrews and Arabs as Lilit, defined in these beliefs as the sister of Satan and whose symbol is the owl or the owl -so used in the occult-, and who is He attributes to her the birth of demons and witches (all of these were called "Lilím", although this voice has sometimes been translated as "mermaids"), from her union with her own brother. This, as Jesus apparently said to John, was what caused the origin of "Destiny": «*He (Sakla) devised a plan with his authorities, which are his powers. Together they raped Sofia and produced something repulsive: Fate, the ultimate, fickle bond. Fate is so because the powers themselves are fickle.*" (Secret Book of John 15:6-7) That is why the Greek texts affirm that from the union of Xaos (Chaos) and Nyx (Night/Lilit) Morus (Fate) arose –in ancient Arab culture fate was also understood as personified, in his case called Nassib, although other times the goddess Manat was spoken of, associated with Alilat by some, while the Egyptians called him Sais and the Taoists called him Bixia Yuanjin. From this point, in Greek mythology, all that arose was the fruit of this Chaos, Darkness, Night and Darkness. Nyx joined Tartarus (Underworld) and had Thanatos (Death), Hypnos (Sleep), Eris

(Discord/War) and other "children". For his part, Erebo, which means "Darkness", was punished by the "deity" and sent to the Abyss, where he became known as "Infernal Shadows". Here may be the point mentioned before, of the deportation of Sakla, Nebruel, Lilit and all their creations (I don't know if the female figure of Lilit and the male figure of Nebruel are, in the long run, one and the same person). In any case, before being sent to the Abyss, Erebo produced Eros (Passion) and Anteros (Rejection, Apathy. Or also "Flower"). Likewise, Erebo/Darkness and Nyx/Night (Lilit) produced Ether (Air and/or High Sky) and Hemera (Day), and less importantly Philotes (Affection. Voice from which the name "pilot" comes). This indicates that they worked together to make their own creation as a replica of the Kingdom of God and trying to put the divine powers into action, materializing them on Earth, in visible and invisible things (the invisible were always first).

« The Mind the God, which is both male and female, and contains in itself Light and Life, gave birth by Name to a second Creative Mind, which, being god of fire and spirit, created in turn 7 governors, container owners of the sensitive cosmos, whose government is called Destiny.» (Treatise XVII, verse 9. Original incomplete and without title. Corpus Hermeticum) Fate was born from the first manifest forces, in practice by Sakla, but by Divine will, without these blind people even realizing it. « *So the administration of the earth is entirely in the hands of the geniuses and is exercised through our bodies. It was this administration that Hermes named Fate .*" (Definitions given by Asclepius to King Amon. Treatise XVI, verse 16. Corpus Hermeticum) The beings that came after the gods were demons and spirits, but certain civilizations did not call them that but rather defined, considered or nicknamed them "geniuses". Through these disembodied forces, the universe is subject to the "Fate" pattern, together with the law of space-time and many other forces that science defines well today. Now,

accommodating the Greek version to the thread of the subject in question, the lowest of the underworld was already defined as Tartarus, without explaining its appearance. But whatever it was that identified that "Abyss" or "Hades" reflected entities personifying forces that we know as death, sleep and discord -aspects that we have already mentioned thanks to other types of literature. For its part, Darkness, which is later sent to the underworld as Infernal Shadows -be careful with the plurality-, before its deportation forms -without any apparent mediation- certain entities, known as Passion and Rejection (although "Anteros" usually reflects in Greek also what we define as "flower", which takes us back to the motor beginning of "flora"), but with its dark consort it gives birth to Ether, Day and Affection. The concept of Ether reflects gas in physics and chemical elements in chemistry, but on a poetic level it personifies Heaven. According to Father Ernetti (a renowned Vatican scientist from the 1970s), the ether would be where each and every one of the external actions undertaken by human beings are collected, that is, each one of us emits millions of waves throughout our lives. life, that they get trapped somewhere.

It is not wrong to say that God made creation in his image, as long as it is also said that everything happened by mistake: both things are right: "*God is always stable, and always, with Him, is Eternity, which constituted, keeping within itself the World that had not yet been born and that, rightly, we call the Sensible World. And in the image of this god this World was made, in imitation of Eternity. Because the World has the vigor and nature of its own stability, even though it is always agitated, due to the very need to return to itself.*" (From Hermes Trismegistus addressed to Asclepius, verse 31. Corpus Hermeticum) However, the Hellenes could be repeating what Moses and others described: the appearance of Heaven and of the Day. as light period, that is, daily illumination. So the darkness preceded the illumination and the negative forces and their energies. Now, around Lilit there are more legends, even in the Sumerian tradition, where it is said that before the reign of Anu, the sovereign of heaven and father of the gods with his companion Antu, Alalu and Lilitu reigned. Alalu was dethroned by Anu – although Anu is also said to have been castrated and later dethroned by Kumarbi, who also castrated him. Some claim that before Anu there were 21 dynasties on the throne of Nibiru, but now Anu had seized power and Alalu was forced to flee to Earth. Anu is identified with the god Uranus, who is called Heaven, a corresponding faculty, being that he is precisely the lord of the gods and patron of heaven, according to the Assyrian, Babylonian, Akkadian and Sumerian culture, without considering that, with the name of Urano/Heaven was understood among the Greco-Romans (it should be noted that the voice Urano has some similarity with the Hindu "Varuna"). Moreover, the story of Kumarbi castrating Anu is the same as the Greek story of Cronus mutilating his father's genitals, which is none other than Uranus, the personification of Heaven. The matter of the names is also discovered with the etymology, since the Roman name "Uranus" owes its root to "Ur-Anu", since the vowels "u" and "o" are equivalent (graphic input does not apply). nor verbally in these mother tongues). The voice "Ur-anu" means "Light of Anu", or literally "light of the sky", "star light" or "light of god", and corresponds, clearly, with the Sumerian An (Anu for the Akkadians), who pictographically was reflected as a star. In fact, this name, being first of the Sumerian numerical order, to the voice "one", in Spanish, or in Greek "ena", to the English "one", since these names could be read from right to left or from left to right. The same with "year" or "annual", as a derivative of "anu", or in line with the beginnings of something: "annals".

In the Niburian concept, from ancient Sumer, it is said that the reptilian goddess Tiamat existed at the beginning, according to some references as the daughter of celestial gods, and there was also "chaos" and darkness from which came the abyss of voids that collapsed. and formed both the Earth and the other bodies of the solar system. Tiamat would be the feminine principle, the sea, representation of the potentialities of pristine chaos. Female monster, malefic in Babylonian legends, the salt water that in union with Apsu (the masculine principle, fresh water) gave birth to gods and animals, but not before starting to create Lahmu and Lahamu (the mangroves). Shortly after creating them, they bothered them with their noise, the god of magic, Ea, managed to subdue Apsu, leaving him in a long slumber (that is why the fresh water would be still) but he could not do anything against Tiamat who Very enraged with the death of her husband, she created a legion of demons led by Kingu, who was her lover and one of her children. The gods decided to give all their powers to Marduk, -son of Enki- he defeated Kingu, who was paralyzed with fear when he saw him arrive, and then Tiamat, who made her mouth open with a gale and shot an arrow inside the stomach. After this, from the blood of Kingu humans were born and from the body of Tiamat, which Marduk chained in the pits of the abyss and split in half, was created, from its upper half the sky and from its lower half the mainland. His tears became the sources of the Tigris and Euphrates (all this legend is parallel with those of Vitra (or Vritra) in Hindu mythology, Cipactli of the Aztec religion, and Typhon in Greek mythology). We do not have to take this account as something literal but metaphorical, since the Earth began as liquid, then muddy and muddy, until the hard earth and dry regions began to be cemented, although Lajemu or Lahmu is dominated or personified by Mars, and Lajaamu or Lahamu for Venus. Apsu and Tiamat only show how the primordial water divided, even resulting in sectors of salty water and sectors of non-salty water, outside on Earth or as an element of the other planets of this system. It is assumed that the greatest potential force in the solar system is the sun, so Apsu is governed by this force greater than him. The demons led by Kingu show that evil forces, entities or spirits appeared that have some close relationship with the initial meaning of Kingu (the Moon in Sumerian).

The meaning of Marduk's incursion has been theorized by several Sumerologists and researchers, as it leads to accepting the idea of an added planet; a foreign visitor part of our system, but that would only be seen every 3,600 years. This intervention would imply the direct action of the rulers of heaven in the modification of terrestrial aspects, since there are gods of heaven and gods of Earth, or in other words, greater gods and lesser gods (the greater ones would never have descended to Earth, since they are the planets of our system or forces that govern them without altering their course or settlement). Later, the reference to Ea (Sumerian god of water, who later became known as Enki and represented exactly the same as the animal alluding to the constellation of Capricorn) suggests the participation of said god of the sea, to whom the preparation is attributed. of the Earth for life, the exploitation of resources and colonization (its resemblance to the Greco-Roman Poseidon/Neptune or the Chinese Yu-Qiang is clear). Said event is referred to in the myth as The Celestial Battle, where Kingu is also narrated, for example, he was a demon who later became Tiamat's second consort, once her first consort, Apsu, died. This allegory supposes that the Primordial Earth ceased to be solely aquatic and passed to a second state, in which its complementary part was associated with the Moon and invisible forces. Tiamat (whose initial word, "ti", means "life" and "love"), convinced that she had to avenge Apsu's death, gave Kingu the "tablets of destiny", these held power and thus turned him into "

Prince of the gods", an aspect that leads one to suppose that the tablets of Destiny, the Me, are the same as the Destiny of Greek cosmogonic terminology. But in the Tiamat contest, he died at the hands of Marduk, and Kingu, being the leader of the rebellion, was sentenced to die since the human race was created with his blood, to serve the gods. The death of Tiamat would represent the geographical division of her "body", splitting, to give rise to the Earth as such. According to a more recent Babylonian myth, all these beings, called Anunnaki, would be children of Anu and Ki, the brother and sister gods, themselves, the offspring of Anshar and Kishar (heavenly pivot and terrestrial pivot, the celestial poles that correspond with Saturn and Jupiter, respectively). Anshar and Kishar, in this other perspective, would be the sons of Lahm and Lahmu ("the muddy ones"), names given to the guards of the temple in Eridu (the Earth already established), the site of Apsu in which the Creation according to they occurred. This novel is referred to in the writing of the story called Enuma Elish, taken from the Ashurbanipal Library (630 BC, although it is said to be 500 years older).

According to the revelation that was given to Noah's firstborn, Shem, Sophia's power materialized, indeed, as a physical property, but wrapped in powers and in a self-determining mind. This is when the light acts that, although it represented the Eternal Kingdoms, was personified in the very generator of the emerging problem. The apparition, in turn, represented a humiliation to the dark being, since it showed not only the light of the Superior Realities, but also the image of the divine being, the Adam code. Although, it can be assumed, from what has been read, that the vision was not sudden and brief, or even unique, but that it also chased the dark one to the low places where he wished to make himself lord, so that the power of the Unconquerable Mind that was in the water, and therefore, in the dark, was returned to herself

– to Sofía. This is how she descended "*so that the mind of darkness could not remain in Hades*" (Paraphrase of Shem 4:5) But why was Sophia so eager to lash out so radically? The reason was "*because the darkness was made, like his mind in a part of the members.*" So, if the Mind of Power was one with the Dark being, the ability to create and develop, in its hand, would have no brake, being one with Nature: «*When, [O] Sem, appeared in it (that is, the resemblance), in order that the dark could become dark to itself, according to the will of the Majesty - in order that the dark could become free of all aspects of energy that it possessed - the mind drew the chaotic fire, with which it was covered, in the midst of darkness and water. And from the darkness of the water it became a cloud, and from the cloud of the womb it took shape. The chaotic fire that was a diversion was there.*" (Paraphrase of Sem 4:10-25) With these words they expose to Sem that the "likeness" of what was revealed was manifested to obscure the dark being -which is none other than Sakla-, since that was the will of the Most High. Then, the Mind, it seems, at that moment, outlined that "chaotic fire" "*with which it was covered.*» How to describe this manifestation with other words? It's like the usual reference to the origin of the world, as a burning mass, I think. It is as if that situation that involved the Mind of power was embodied outside of it, specifically between the "*darkness and the water*" itself. Could that powerful Mind be invaded by Sakla's negative quality? Is what it seems. Removing the chaotic thought from herself caused "the tohú and bohú" in the Darkness and the water, precisely what Sofía was preparing to distance, separating the elements, powers, beings and forces. Now, it can be glimpsed that this Mind came to be transformed into a kind of "cloud" difficult to describe, but in whose womb something began to take shape: the chaotic fire. Ergo, we can find that these definitions of cloud, chaos, water, darkness and fire are the ones that always appear in mythological stories.

The Primal Waters

As in the Mediterranean novels, the Norse recorded the appearance of more things from Chaos: Narvi had Nott (Night), and she was married three times. In each marriage she had a son: from her first husband, Naglfari, she had Aud; from the second, with Annar, he had Earth or Erd (Earth); from the third, with Delling (the Twilight), she had Dagur (the Day). The name "Naglfari", part of Naga (from Sanskrit root "snake"), Aud part of "Iud", base to name the supreme deity according to the Semitic origin of that voice. Annar reminds us of the Niburian Annu, Anu or An, whom, according to a myth, Alalu, Lilitu's consort, castrated and dethroned. It is, in fact, a completely Niburian name (term used by Zitchin to identify the Anunnaki). Dagur is also Sumerian, and its Hebrew root comes from "Dagar" (the fish), from which the Canaanites got the name "Dagon" (Dragon). In the Teutonic and Scandinavian culture, the story was very different, although inside it had reminiscences similar to the typical story. In it, yes, there were giants, and as in Greek and Egyptian mythology, a primordial cow (which may be related to the fact that the galaxy is defined as the Milky Way, as in the case of Taoism, where it is understood as a river called Han -literally "river"-, or more specifically "River of Heaven", "Silver Han" or "Han of the Stars"), but the appearance of the universe arose by another type of demiurge, a giant whose parts created the 9 existing worlds or kingdoms (it should be noted that also in Taoism it is believed in the existence of 9 kingdoms). One of them, Muspell (or Mushpels), the hot land, melted the frozen land of Niflheim. From said melting ice arose the cow Audhumla, which licked the salt blocks of Ginnagagap and gave rise to Ymir, the father of the first race of giants. This cow also produced Buri, father of Bor, who in turn was the father of the three gods of the Aesir and rulers of Asgard: Odin, Vili and Ve. We find striking points here: Audhumla, the cow, recalls the Egyptian beef Mehturt, which symbolizes the Milky Way, as well as Amalthea, the ruminant that apparently suckled Zeus in his infancy. I add that Mehturt, whose epithet is "Mother of Ra", means "Great Flood".

It should be noted that Muspell (Mushpellr or Muspelheim) is similar to the Hebrew "Mishpat" (trial or punishment) or "Mishpaja" (clans or families), and that the word Niflheim, like the rest of the kingdoms, uses the Hebrew plural. Niflheim seems etymologically similar to "Nefilhim" or "Ha-Nefilim" (the Fallen), a word widely used to name a famous race of giants, "those who *fell from heaven to Earth* ", which in Sumerian would sound like: "ANNU.NNA.KI" and in Hebrew is "Anakim" (the sons of Anac, the giant). The name Ymir is similar to "Yomir", whose root is "of the day" or "of the aeon". Similarly, the name of the Ginnagagap columns is surprising: "Ginna" comes from "Gin", and this from the Arabic "Jinn" ("genius" or "demon"), is also associated with the Greek Hecatonchierus, son of Uranus (Sky) and Gaia/Gea (Earth), called Giges, and also reminds of the Hebrew "gan" (garden or orchard). Although Gyges was one of the three Hecatonymichirs (those with 100 arms) from which the voice "Giant" comes out. Even the ending of the name Ginnagagap is significant: "naga" and "gagap" have an Indo-European source. The Naga were respected divinities in Ethiopia, to the extent that there is a book called "Kebra Negast", the book of Rastafarian wisdom and faith of Ethiopia and Jamaica. The name of the book is similar to "Cobra Najash". Nagas in Indian culture were serpent gods, and this word passed to the Hebrew "Najash" (serpent). But this point does not end there: "Gagap" could easily come from the Sumerian "gaga" -like the singer Lady Gaga-, which was the name of the planet Pluto, which the Romans identified with Hades. Also "agaga" comes from "igigi" (the Anunnaki astronauts of the Sumerian civilization), who are considered to have descended to Earth and had children with the daughters of men. Norse culture may be more information-rich than meets the eye. The very word "igigi" is the transformation into Greek that is said "Grigori" (watcher), a term that the translators of the book of Enoch used to transform the Hebrew plural word "irím" (observers), since it called the fallen angels of Mount Zion (also known as Hermon).

Furthermore, the Norse story specifies that Ginnagagap or Ginnungagap means: "Resembles the Void", and was understood as a primordial void that separated Niflheim and Muspell, the land of eternal ice and the land of eternal heat and flame. Before there were men or gods, the Hvergelmir Spring, deep in the lost ice of Niflheim, began to give rise to 12 rivers known as Elivagar. The Audhumla cow licked from one of these blocks of salt, later dried. According to the Asatrú tradition (Viking Eddas) it is related: « *When neither the Earth nor the sea nor the air existed yet, when only darkness existed, the Allfather (Father of All) was already there. At the beginning of creation, in the very center of space opened Ginnunga, the terrible bottomless and lightless abyss; to their north was the Land of Niflheim, a world of water and darkness that opened around the eternal spring of Hvergelmir, source of the twelve rivers of Elivagar, the twelve streams that ran to the edge of their world, once of encountering the wall of cold that froze its waters, making it also fall into the central abyss with a deafening roar.* » It is interesting to see that Norse mythology says that the beginnings of the world came as a result of two great kingdoms totally opposite in reality. The recent Thor film locates the 9 kingdoms in places not far from each other, in the middle of the cosmos, connected to each other in what would be a great visual tree of plasma and gases in colossal proportions. That idea that the production of the film had could also be considered, knowing that Earth is one of several worlds existing in our neighborhood. The story also obeys an elementary principle: God, "the Father of All", or "of All", who was later confused with Odin. We could rightly say that this part of the narrative does not differ from those already mentioned.

In the root Varna we find that, according to the ancient Vedas of India, one of the legends says that Creation is the result of the sacrifice of Purusha (Man), the primeval being that is all that exists, understood in "whatever has been *and how much will it be.*» When the sacrifice of Purusha, who had "*a thousand heads, a thousand eyes and a thousand feet* ", was consummated, the clear butter that was formed became the animals that live on Earth. From this same sacrifice were born the gods, Indra (Menacing King), Agni (Fire) and Vayu (Wind), as well as the Sun and the Moon. The atmosphere was formed from the navel of Purusha, from his head Paradise arose (it is clear that the concept of Paradise was already considered in the peoples of the Far East), from his feet the Earth, from his ears Heaven. Another narration affirms that Vritra, an ancient serpent (known as the celestial dragon or first dragon) -equivalent to the Egyptian Apep-, was defeated by Indra, with the help of Vishnu (Vishnu), and at that moment that god (Indra) created the sun, the sky and the dawn, while the body of the dragon flowed the waters of the Earth that had been retained. In relation to Vishnu, it is said that he helped to the explosion of the universe and to separate Heaven and Earth, presumably in line with this he is considered as the "cosmic column". The most interesting world about the history of Vishnu/Vishnu, is not himself, but his consort, Lakshmi, whose name reminds us of the "muddy" of the Mesopotamian culture (Lakhmu and Lahamu), not only because of the linguistic parallelism but also because of their interaction and essence: Lakshmi is known to be one of the things or creatures that came out when the ocean churned. Hinduism conceives the origin of being as a primordial ocean, called Prajapati or Sarasvati, sometimes deified. Various gods or powers (called "asu"), sacred malignant ("asuras") or benign ("devas") would have emerged from that elemental bosom. Other forces or primordial entities are also defined as the "rudras" spirits.

Observing the teachings of Hermes Trismegistus (whom he assumes is the grandson of Hermes/Mercury, possibly the messenger of the Greek gods), he identifies Earth as the "offspring" of the Most High God, suggesting an association between the life generated by God He manifested himself by bringing this planet into existence as his "son", representing in himself the man not yet engendered, but which would encompass man, reflected in the multiplicity of souls that would incarnate here. But in much later texts from India, known as the Puranas, the story of Creation is more complicated: the Creator of the Universe was the god Brahma, who had arisen from the Primeval Ocean and existed by himself (Swayambhu). Brahma transformed into a huge boar (Varaha) to raise the Earth from the depths of the ocean, something quite similar to the myths of the appearance of the world according to Mayan, Aztec and Gnostic texts. Another scripture says that the cosmos was released or arose from Hiranygagarha (discuss the phonetic and compound similarity to the Norse "Ginnagagap"), the "Golden Matrix", which was the Primordial Ocean, which was understood as a Water of the Void and the Darkness of Non-Existence. With certain nuances similar to the Chinese myth, from the "Golden Matrix" a "Golden Egg" was released that formed the universe, Swarga and Pritviti (the Earth. Word whose root could reach the Semitic "Prit": "Fruit"). This great beginning is referred to, in an Egyptian version, as an infinite sea that was in the beginning. It would be a place without life and in absolute silence. Then Ptah would appear -who could also be related to Sakla- with the "forms" of the abysses and the "distances", of the solitudes and of the forces. His "thought" would have taken the form of Atum and, devouring his own seed, would give birth to Shu (Wind or Air) and Tefnut (Moisture), whom he expelled from his "mouth", thus creating Nut (Heaven) and Geb (Earth). It is said that in another version those 9 principles were before Ptah and then it would have been when he appeared conceiving from himself and then he would "rest". The popular myth says that Ptah created the Sky as a conductor and surrounded the Earth with sea. This demiurge would also have created Tartarus to appease the dead. Then, after that, he would have placed Ra over Creation, Ra coming to give limits to night and day, and to fix their seasons. In error, once again seen, in the peoples of yesteryear, it is also seen reflected here, and I add in this regard that it was assumed that Ra subdued the forces of darkness. Ra was associated with Amun, since a legend said that Amun (Amen, Ammon or Amun, confused with another figure whose name resembles) appeared from an "egg", which when broken in a flash gave rise to the stars and other luminaries (the tale varies depending on the New or Old Empire).

Parallel to the Greek principles, this cosmogony is manifested in the pantheon of their North African neighbors, the Egyptians, who say that Ptah was the creator god -especially in Memphis- although Ra was the creator god on Earth -Ptah would be in the darling. He is seen as a kind of "creator", from which emerged the world, its inhabitants and the Kas (spirits) of the other gods (both Egyptian and Hindu gods -even Taoists- have the peculiarity of being portrayed with green, blue or violet, to identify them as such and/or for some other reason of remote nuances). It is difficult to trace the origin of Ptah in the Ennead of Heliopolis, but he seems to be the creator who gave rise to the demiurge Atum and the creator sun god Ra, as ruler of the Earth (although he was rarely on it). Ptah is said to have been the first ruler of Earth -which is why the Sumerologist Zecharia Sitchin associated him with the Sumerian/Akkadian Ea/Enki-, voluntarily abandoning this work and leaving it to his son Ra, who, in due course he did the same with Osiris, and Osiris with Horus. But something striking is the other name of Ptah (Ta-Tenen or Tanen) as god of the Earth, since it resembles the Hebrew voice "Tanín" or "Ha-Tanín" (the Dragon), being that precisely Ptah is describes him as a mummified man with "green skin". Ptah is said to have been the father, with his wife Sakhmet and Bastet, of Imhotep (the famous architect usually associated with the Greek Asclepius/Aesculapius), Nefertem and Maahes. Notice again a parallel with "Nephilim" (NFM Hebrew, and its mother tongues, do not have vowels), in this case in relation to "Nefertem". His mother, Sakhmet, was precisely the goddess of revenge, later baptized as Hathor. She was sent by Ra to destroy the Earth and almost succeeded. She was known for causing epidemics, so she was called "Miss Pestilence." She was identified as a woman with the head of a lioness. If Ra comes from Ptah, this is complicated, as another Egyptian version says that Ra (the personification of the sun), created himself from the Primal Waters of Nun or from the primordial lotus flower (a parallel with the Greek Anteros, the Flower, although the lotus flower is potentially relevant in Eastern culture, especially in relation to the gods). Ptah has sometimes been associated with the Greek Typhon, as is the case with Seth, and these associations sometimes confuse scholars more than Egyptologists themselves. Then we have that Amun appears as lord of Thebes (Jeremiah 46:25) and takes the place as an important deity between the Greeks (Ammon) and the Egyptians (Amon-Ra), replacing Ra, for which reason Sitchin linked him with the Babylonian Marduk. In fact, it becomes known in Canaan as "Baal." His companions are said to have been Amaunet, Ipet, Mut, and Nut/Bast. From his union with Mut (mother. A vulture goddess) he had Chons (god of the moon).

Nut or Niut, was the goddess of the sky (in English "Sky") and the heavens (in English "Heaven"), and her official consort was Geb (Earth). Another Egyptian version says that Ra came from

Nun, reporting that "*NUN continued to create... the sky, the air, plants, animals and gods, but something was missing, there was no absolute darkness, but there was no light either. One day, from a lotus floating on the Nile, light arose. The flower refused to open and when it couldn't take it anymore, RA, the sun, was born from inside, giving the world what it was missing, that light with which to appreciate the colors, the beauty of creation and of course time, since RA returned to the interior of the chalice of the lotus flower to rest while the night lasted. RA became the most powerful god, the master of the world and also the most envied ...*" We must point out the characteristic attribute of Ra as an "eye", which is usually seen outside of him identifying himself in other gods or goddesses, especially in Horus. Precisely the voice "Re" or "Ra" is Mesopotamian, and comes from "See", which explains the meaning of the eye and the concept of: "*the one who sees everything*". Ra, in any case, was seen from time to time in his "Aton", a kind of parallelism with the Bennu or Phoenix that from time to time came down from the sky to the place of the Sun Stone, or landing platform, like the which was discovered in Baal-Bek, Lebanon. Tradition says that Ra (whom the Greeks associate with Helios, the Greek sun god, son of Hyperion and Trya/Tia, who were born to Uranus, the sky, and Gaia, the Earth, the first giants, who came from Ether and Hemera, and Eros and Nyx, respectively) was the father of Bastet (goddess with the head of a cat), of Hathor (woman with the head of a cow. Associated with Astarte/Aphrodite) and of Sia and Hu, the personifications of wisdom, which would have arisen from drops of blood from his penis. It is also worth remembering that the Greeks considered that Cronos (Time) castrated his father Urano, and his phallus fell into the sea. From the foam that arose from there, Aphrodite was born, who has always been related to beauty, love and passion (Eros). However, the most significant children attributed to Ra/Re, due to his masturbation, are Shu (Air) and Tefnut (Moisture), which links them to the Greek Ether, the son of Erebo and Nyx.

In Egyptian culture, Atum, the personification of the intraterrestrial sun, was the father of the gods and was already diluted in the waters of Nun, manifesting as the first solid matter that emerged. According to the myth, at the beginning it was found in said primordial ocean in the form of a serpent. This character is also reminiscent of the story of Sakla, as the legend adds that seeing himself alone he decided to create the first Heliopolitan ennead, and his own representation is important in this regard: "man with the head of a ram" (let us take into account that the visions of the prophet Daniel in Babylon they seemed to identify the struggles of all these potentates and authorities, and there some of these beings are identified under very characteristic symbols such as the ram or the male goat). From here we go on to the sons of Shu and Tefnut, to name their two sons: Geb (Gueb or Seb = Earth) and Nut (Nuit = Heaven). Note the parallelism between the goddess "Nuit" and the name of the night in French. It is precisely Nuit/Nut who is said to have played an important role in "*separating the forces of Chaos from the order of the Cosmos*" in this world. It is said that once she "*formed the firmament*" over her husband Geb/Seb (the Earth), which in Aramaic means "pit" or "back" (in Hebrew, Geb, can mean: beam, cistern or locust). The very name followed by her husband, Seb, also has etymological roots that lead one to the names "wolf" and "flying" (the name can be written as Geb or Seb). It has great parallelism with the Chinese deity called Pan-Gu (see linguistic parallelism with the Greek "Pan-Gea", name of the terrestrial super-continent), who was born from the "cosmic egg" as a dwarf and became a giant by separating the Heaven from Earth (another myth assumes the creation of Heaven and Earth, and all that has life in them, to the male deities Mu-Gong, and female Xi Wang-Mu), similar origin of Sun Wu-Kung (Wildlife Monkey), famous in Chinese traditions and known as the Monkey King. It is said that Sun Wu-Kung was born from the "Primal Chaos", and emerged from a stone egg that fell from Heaven - like the Egyptian tale of Amun. In the classical Chinese concept, it is said that there was a primordial principle, a chaos from which The Venerables that formed Heaven and Earth emerged, the same ones that would rule from Heaven. Now, in Mesoamerica this tradition was similar: the Aztecs or Mexicas said that everything arose from an Immense Sea where the Earth monster, Cipactli, lived. This gigantic alligator was killed by the brothers Quetzatcóatl and Tezcatlipoca, and with its parts they formed the world. The origin of the deity is similar to a monotheism mixed with two concepts of duality: Ometéolt and Omecíhuatl, the parents of the first 4 gods (Tezcatlipoca, Quetzatcóatl, Tláloc and Ehécatl).

Centuries before, the Mayans had already had these ideas, affirming that everything arose from the alligator-goddess, Tlaltecuhtli, a sea monster that was Lord of the Earth, and which was deceived by Kukulcán (the Aztec Quetzatcóatl) and Tepeu to capture him and stretch him/her, forming the Earth from him/her. We could relate this beginning as a beginning where the light "exploded". Illumination arose from everything that was not seen and God manifested what he wanted to create. The next thing was the appearance of solid things. It seems that here it was identified that the light is what is above everything and there is nothing above it, and the lower parts below which there is nothing either. In matter that does not make sense, but it can if we talk about the "heavens" and the "dimensions" or if we identify this event with the geophysical division that would create dry areas and aquiferous areas, as well as other planets. It would be understood that the highest in the heavens of the Most High is light, and below the lower level of creation -where the Abyss is- there is nothing because it is the lowest: darkness.

For this reason it is good to quote a phrase from the book of Job, which says: " *He limited a circle on the face of the waters, until the end of light with darkness.*" (Job 26:10. Hebrew version) It seems that from there, and with light, God made water appear, a kind of universal ocean, or simply the particles of a water that filled everything and that were scattered throughout the cosmos like part of Creation, which in the Nag Hammadi texts is attributed to the error committed by Wisdom – which, in the long run, would not cease to be the design of the universal Father and, by extension, his will carried out, as always. On the other hand, it would simply be a compound with great potential, which would be suspended in space, and with which other bodies would be formed, but not before removing the bad and toxic elements that were mixed there. Certainly the universe is so vast that, if so, a powerful hand could gather water particles from everywhere and even assemble watery planets. It is known today that the universe is charged with energy particles and basic molecules of living matter, which have been found to be "floating" in interplanetary space. Obviously these discoveries shattered the belief that life could only exist within a certain range of atmospheres or temperatures, so it is preferable not to give it much echo. In any case, having made a quick and unbridled tour of the most striking mythologies of our globe, we have observed how created things received names and personifications as deities, but still, we will not go into more detail that they exist without first going back a little, to where Sofia tried to work things out. From when all the observable creation took place, the records began to document, but all the previous history was missing, the one that explains how the visible aspects themselves work. If we then look in that direction, before the origin of these gods, we will understand that the entire cosmos is pure energy, pure power and existing forces interrelated with everything. For this reason, since ancient times, man, gods and angels have been linked to stars. Now, Sofia gives rise to a mass of dark water, chaos, and then tries to correct what got out of hand, intervening and trying to distance this immensity from everything that is light and dwells in the imperishable. This leads to the mythical legend of the separation of heaven and Earth, of the waters, one from the other, of the visible and invisible, of the liquid and dry, and of heaven and the underworld, understanding that the maximum personification of darkness it is Sakla.

The Abyss

The legendary prophet wrote: «*saw great rivers and I came to a great darkness and to where no carnal being walks. I saw the mountains of winter darkness and the place where all the waters of the deep flow. And I saw the mouth of all the rivers of the earth and the mouth of the deep.*" (1st Enoch 17:6-8) As in the ancient narratives, in Genesis it seems that they begin to reveal to us that the first things were already "Tohú ve Bohú" (chaos and darkness or desolation). So, before the light, there was already darkness? It depends what we mean by light and what we mean by darkness. After the over and over again scene from the first book of Moses, it is said: "*Let there be light.*" Clearly light makes "invisible things" visible things. Thanks to light we can see shapes and colors, everything under the spectrum or level of light intensity: there are different fields of light -and energy-, such as infrared, visible light and ultraviolet. Without the bounce or return of the impact of light against an object and its return to the receptor (the eye) we would not see anything. For this reason we could understand Scripture, where it says that "*By faith we understand that the universe was constituted by the word of God, so that what is seen was made of what was not seen.*" (Hebrews 11:3) Initially, great things were done and then lighting appeared so that we could see it, although that does not mean that the light that appeared in said story referred to the divine appearance, destined to eliminate the mass or power of the created darkness. But, did we live in a universe that "was not seen" before "was light"? We are talking about the spiritual creating the mental, and the mental creating the physical, so we are talking about a story that has already begun whose beginning we have begun to unearth. Without going too far, dealing solely with the biblical source, we find that, according to the 1960 Reina Valera translation, "*there was darkness on the face of the deep.* » That is, things already created, with which it was not the beginning but what we see now. That millennial abyss was "covered" or "lined over" with mass of firm rock, thus distancing it from the surface: "*From the beginning you founded the earth, And the heavens are the work of your hands.*" (Psalm 102:25 and Hebrews 1:10) Did God give the Earth a "foundation"? That is to say, on a formless, dark, invisible and/or immaterial place he "made hard rock" –as the prophet Enoch reiterates. The Earth seems then to be "founded" on "columns" or "supports" that were not there before, and that apparently came out of an ancient and eternal sea: «*He founded the earth on its foundations; It will never be removed.*" (Psalm 104:5, 82:5 and 18:7, Isaiah 24:18 and 51:16 and Micah 6:2)

The theory of terrestrial terra-formation is neither new nor exclusive to ancient mythologies, since it is said that this globe was built as a concave body that would "differentiate" the Abyss, under the layer of earth - and that it would now be its "covering" – and the sky, or universe, from each other. Were they trying to isolate the Abyss? For what purpose? Let us bear in mind that the Hollow Earth Theory does not apply only to our planet but to all celestial bodies, which complicates the idea of the exclusivity of our world in many aspects, but, even with everything, it does not deny that this orb hold control over dark forces that are isolated below it, and at the same time fulfill a transcendental role in the history of the cosmos. In addition to this, there is already talk of "waters", and about this there are more hypotheses: some literally suggest that water already existed; others affirm that the water is *"nations, peoples, crowds and languages"* (reference Revelation 17:15); others suggest that the composition of water (2 molecules of Hydrogen and 1 molecule of Oxygen) was broken up by the universe and was *" gathered in one place "*, creating what we call "water" (this would explain why there is water above the sky). : Psalm 148:4) and seas, while another maintains that the Earth was already there from the beginning but it was a gigantic mass of water, which the Sumerians would have called "Tiamat" (TIAMAT. Precisely a similar voice -and from the same root- to the Hebrew "TEHEMOT": "Abyss"). In any case, let us not forget that in this sense we are talking about a single plane, since Set, Enoch, Baruch, Isaiah, Pablo and Juan in their annexed texts, very possibly spoke of more than 3 existential planes or heavens -specifically up to 7 some times, and others up to 10-, which do not belong to this dimension or simply to a known location, and it would be necessary to inquire deeply into the origin of this Abyss. What Enoch, Moses and Zecharia Sitchin seem to explain –who claims to get his information from Sumerian tablets- is a separation of the existing body of water. Other theories speak of "distancing" between groups that remained on Earth and in heavenly abodes after the preconceived idea of the "satanic rebellion", and to whom a specific location had not been predetermined. The most important points that the researcher and scholar of Genesis must attend to are the references to said masses of "water", the so often repeated "abyss" and the misunderstood "heavens", because they play the decisive role in the vision of how it was built the Earth and simultaneously other planets.

If apparently an Abyss existed in a "universe" or "dimension" that to the physical eye was not, or is, perceptible, or simply symbolized a gloomy hollow, there would also be other imperishable "abodes" or "worlds" that came with "the Word". Then "light" was made so that everything invisible in the cosmos would be visible, or to clarify what was happening, clearing away the evil, or indeed the light came with Wisdom to -either symbolically or literally- solve the problem with the darkness appeared Here we see that this Abyss seems to have been located within the "created earth", where first there was "water" and later "solid rock". The Scripture says in this regard: « *These are voluntarily ignorant, that in ancient times the heavens were made by the word of God, and also the earth, which <u>comes from water</u> and by water subsists, for which reason the world of then perished drowned in water ...*" (2nd Peter 3:5-6) In other words, that the Earth "comes from water", as we see today, that there are planets that are immense stores of minerals and compounds from the Periodic Table of Elements, and Of course, among them is water in various forms, but present in all of them –although in sight, or officially, it is not exposed. Let us evaluate this by remembering the reports that today NASA tells fables about planets with water like ours (I say "fables" because NASA knows that there are planets like ours, but nevertheless they have hidden the truth from us for decades). In this case, would Tiamat be a gigantic body of water? There are many possible theories, but what is clear is that the Abyss was there from the beginning and it was an area that had to be limited: «*Before the abysses I was generated; When there were no fountains abounding with water.*» (Proverbs 8:24, Psalms 33:7, 71:20 and 135:6-7). The source is the origin of something, in this case, it is possible that the water. We also find that it says: "*By his knowledge the abysses were divided, And the heavens distilled dew.*" (Proverbs 3:20). And likewise we find: "*He set a limit to the surface of the waters, To the end of light and darkness.*" (Job 26:10) It is clear that the Abyss was to be limited and heaven separated from "*those who were in the abyss*", or those who would be deported to the Abyss: "*When I formed the heavens, there I was; When he traced the circle on the face of the abyss;*

When he affirmed the heavens above, when he affirmed the sources of the abyss ..." (Proverbs 8:27-28) Similar words come from the book of the Edomite Job: "*Sheol is naked before Him, and Abaddon has no cover.*" (Job 26:6. Hebrew version)

In the Book of Jubilees, its author wrote that in the First Aeon (Interval, Day, Century, Age) Jehovah Elohim did " *seven great works* ", specifically: the creation of the SHAMAIM (in Akkadian: "People of the Rocket", and in Hebrew: "There are Peoples there"), translated as "heavens", which is above the Earth; the second would be the MAIM (waters); then all the wind-spirits that serve the Lord, that is, the angels; then the spirits of the rest of the creatures that are in the SHAMAIM and those that are in the ARETZ (Earth); the TEHOM (Abyss) of the JOSHEJ (Darkness) and the LAILA (Night); finally, the OR (Light), the EREB or SHAJAR (Dawn) and the IOM (Day [of light]). This leads us to think that the Earth could originate from an already created beginning, where there was considerable water to fill a planet like ours, and also there was a deep interest in separating the "lower" part from the part that later remained Above. Let's not forget in any case that the Earth's atmosphere -apparently-, before the Flood, was a dense layer of water in suspended particles that "teared off", but we will mention this later. Well, if during the formation of the solar system, including the Earth, there was an immeasurable liquid body that was "divided" to create the Earth, where did the rest of the water on top go? In August 1989, the unmanned spacecraft designated "Voyager 2" flew far toward Neptune and demonstrated what it really was like: Neptune is blue-green, largely made up of water, and has spots whose coloration resembles "vegetation." of the swamps." This last tantalizing aspect may denote more of a color code if the implications of the discoveries from its satellite, Triton, are taken into account: there, "*dark broads with bright halos*" have suggested to NASA scientists the existence of "*deep pools of organic mud."* We can see the same in Uranus, since in January 1986 its most striking characteristic manifested itself: its blue-green color, a totally different color from that of all previously seen planets. Challenging what scientists have postulated, Voyager 2 showed that this is a being covered not only by a layer of frozen ice on its surface, but also by an ocean of water. In other words, a gaseous atmosphere was found, which in fact envelops the planet, but beneath it an immense layer of about 6,000 kilometers thick, of *"superheated water, [whose] temperature [is] as high as 8000° Fahrenheit "* (in the words of JPL analysts). Even more fascinatingly, the appearance of the Earth seems to accompany the appearance of the last "giants" of the solar system: Uranus and Neptune.

These, as well as the mysterious atmospheres of Venus and Mars, point to the possibility that huge astral bodies existed, from which the planets of our solar system – or one in particular – were formed. Clearly they could not have developed the principles of physical laws in and of themselves, nor would they have given rise to the many capricious, teeming life forms. Today it is known that Mars had oceans thousands of years ago, which either absorbed the surface, filling underground deposits in its "hollow core" or in galleries, and/or evaporated into the atmosphere, creating the typical Mars scene with a bark reminiscent of a constant greenhouse effect. Who knows? It is possible that this past was not very far from that of Venus – although contact groups say that life thrives under its dense atmosphere, and it even has vast oceans like ours. If the Sumerians were right, and Sitchin knew how to correctly interpret the tablets of the Enuma Elish (Sumerian Epic of Creation), we would be accepting that at the beginning there were at least 1 to 3 large bodies, *"when neither the sun nor the moon were yet»*, and when the largest or most influential one entered its elliptical trajectory, it attracted the "ancient" already residents of this primitive system, causing a chain reaction of impacts that fragmented Tiamat and other "neighbors" or "travellers" (although the other theories would be that another "effect" caused the separation). If our solar system were a battlefield of titans, we would understand the allegories of so many legends about the celestial battles, referred to as radical planetary changes, motivated by laws, powers and forces of the mental universe, to isolate those who appeared in the dark: «*I left my body, a bright flame in the night. I stood before the LORD, bathed in the fire of LIFE. Taken I was then by a force, far beyond the knowledge of man. Thrown I went into the Abyss through spaces unknown to man. I saw the molds of Order from the chaos and the angles of the night. I saw the LIGHT, blossoming from Order and heard the voice of the Light. I saw the flame of the Abyss, thrown from Order and Light. I saw Order emerge from chaos. I saw the Light giving Life. Then I heard the voice: Listen and understand. The flame is the source of all things, potentially containing all things within it. The Order that sent light is the WORD and from the WORD, LIFE EMERGES and the existence of everything.*" (Table 9. Emerald Tablets of Thoth)

Let's note another fact: mythologies regularly assume that "a body" that was in the "primal water" – or was composed of the primeval water – was divided and created Heaven and Earth. As I said before, specifically the name "heaven" in Greek is "uranus". It is as if a "giant" had been split in half and its two most important parts created Earth and Uranus, or something similar. Another detail is that Neptune represents precisely the "god of the sea", Poseidon, although how were the Greeks to know that Neptune had water? In the same way, Pluto is the Roman name of the god Hades, the lord of the underworld, of the "Abyss", as we see in astronomy: Pluto is theoretically the furthest planet in the system and is "plunged into darkness" -although Sedna, and other planetoids beyond the aforementioned, have recently been discovered between the Kuiper belt and the Oort cloud. But let's not go too far, since Venus, the beautiful star, was associated with Aphrodite, and red Mars with the warlord, Ares. In the Bronze Age, how could they tell the color of the planets? And how did they know that there was a decisive "war" on Mars in the prehistory of our solar system? There is so much to understand about cosmogony and the way the ancients tried to expound what they had been taught that we would have to get inside their heads and then struggle to understand, with our current vocabulary and advances, what they might have meant. It is obvious that material that narrated events prior to the appearance of said primordial water, chaos, darkness and the abyss, did not come to light until very recently. So, with what has been given to us today, we can study the past of the gods and the beginning of the companions of the universal Father. An imperishable messenger told Shem that after the excellent eternal light something unheard appeared: "*And the darkness was the spirit in the waters. His mind was wrapped in a chaotic fire. And the Spirit between them was a soft and humble light. These are the three roots. They each reigned in themselves, alone. And they covered each other, each with his power.*» (The Paraphrase of Shem 2:1-10)

The definition of "Spirit" refers to many things, but usually in association with Genesis 1, the one who "*hovered over the face of the waters*", is used to speak of Wisdom, trying to fix the disorder: "*Then the Mother began to move from one side to the other. She realized she was missing something when the brightness of the light dimmed. It became dark because her lover had not collaborated with her. I said, "Lord, what does it mean that she moved from one side to the other?" The Lord laughed and said: "Do not suppose that it happened just as Moses said, 'on the waters.' No, when she recognized the evil that had taken place and the theft that her son had committed, she repented. Although in the darkness he had forgotten his ignorance, he began to feel ashamed and agitated. This agitation is the movement from one side to the other." "The Arrogant took power from his Mother. He was ignorant, for he believed that no other power existed except his Mother."* (Secret Book of John 8:1-5) According to Mesopotamian mythology, the universe appeared when Nammu, a formless abyss, opened itself up and, in an act of self-procreation, gave birth to An (sky god) and Ki (goddess of the Earth). In those principles would also be Ab-Zu, or Apsu in Akkadian, the progenitor of the planets of our solar system. The name was then adopted to Enki's abode in Eridu. Later, after the Deluge, Apsu took the connotation of "The Deep" (in Greek "Abyssos") or underworld, although some suggest that it referred to the mines of South Africa. The process of understanding how Abzu/Apsu went from being a giant deity from the far reaches of the galaxy to becoming the underworld can be rooted in everything we have previously discussed. Precisely the change in the meaning of Abzu over time does not imply the influence of the peoples of the Earth or the deterioration of the culture, but the evident process to which the power of Sofia, Sakla and their gods was plunged. The part or idea that associates Absu with South Africa may be due to several reasons, one of which could be the linkage of the "down" aspects, referring to an underground site, such as the designation of an area "south" of the known land. It should be noted that speaking of the "south" was a synonym or parallelism with "below", only changing the perspective according to the viewer or analyst. At the same time, South Africa, at the time of the Anunnaki (space visitors who established the Sumerian civilization), was founded as a mining center, carrying out speleology deep in the bowels of the Earth, including galleries or pits, all with the in order to exploit mining. Apsu recalls the Greek Erebus who, beginning as a pre-titan engine, was forced into exile in the lower parts of our world, as Zeus later did with his brother Hades, and many others, appointing gods and demigods to the custody of the underworld.

In Hebrew also understood as Abaddon, the abyss is a concept and a place that contains many mysteries: «*I thought of the produced Ennoias that came out of the Spirit without stain, on the descent into the water, that is, the regions below. And they were all of one mind, as it is out of one.*" (2nd Treaty of the Great Set) The abyss is the word that describes a DEPTH, the abyss of a cliff or the deepest part of a great hole, sometimes visualized as a fall without limits. It is what pulls down the most and can be related to the Hebrew name "Abadon" (in Greek it is Apollyon, that is: "DESTRUCTION"). Such an approach is explained in the Revelation of John: " *And they have for their king over them the angel of the abyss, whose name in Hebrew is Abaddon, and in Greek, Apollyon.*" (Revelation 9:11) Abaddon is not the name of the angel, as has been misunderstood, but the name of the Abyss. When speaking of "abyss" reference is also made to the "ocean depths", almost always to the depths of the waters or the sea. In essence we see this when he says about Israel, when crossing the red sea: « *... the one who led them through the abysses, like a horse through the desert, without stumbling?*» (Isaiah 63:13) The term "Abadon" has several meanings: annihilation, destruction, ruin and extermination. Its root "Abdá" (loss or disappearance) gives rise to "Abad", which means: to stray, lose oneself, err, lose, wander, perish, be destroyed or disappear. Abad can also be associated with "slave", and may have a resemblance to "Obed", which is "worker". For its part, Hades or Sheol is the underworld, ultraworld, subterranean world, afterlife, world of the dead, and is normally identified as a place next to Abaddon (Abyss/Apollyon). Definitely the antagonistic site of heaven are the abysses: "*They ascend to the heavens, they descend into the abysses ...*" (Psalm 107:26) Apparently the psalmist speaks of God resurrecting him on the last day, or perhaps he was speaking of the Anointed One: «*You, who have made me see many anguish and evils, You will give me life again, And again you will raise me from the abysses of the earth.*» (Psalm 71:20) It may not refer to the fact that the psalmist is in the Abyss, but that God will restore him by bringing him from down there, from the depths. Even Ezekiel wrote: «*And I will bring you down with those who go down to the grave, with the peoples of other centuries, and I will put you in the depths of the earth, like the ruins of old, with those who go down to the grave, so that you will never again be populated; and I will give glory in the land of the living.*" (Ezekiel 26:20) That place was identified by Job as " *a land of darkness and the shadow of death.*" (Job 10:21) Another example says: « *Silver certainly has its springs, and gold a place where it is refined. Iron is taken from dust, and copper is melted from stone. They put an end to the darkness, And they examine everything to perfection, The stones that are in darkness and in the shadow of death. They open mines far from the inhabited, In forgotten places, where the foot does not pass. They are suspended and balanced, away from the other men.*» (Job 28:1-4)

This is significant because the "ancient deserts" -as RVA 60 translates-, or "peoples of yesteryear", have been covered by sand, basically in the Middle East. Another example is also taken from the book of Jobab, where it is stated: « *If my heart was deceived by some woman, if I was lying in wait at my neighbor's door, let my wife grind for another and others bow down on her! Because that is wickedness and iniquity that the judges have to punish. Because that is a fire that would devour even Abaddon and consume all my property.*" (Job 31:9-12) If the Abyss identifies the depths, the theory of the "abysses of waters", from which our planet was formed, could have more argument: «*I was lying in the house of my grandfather Mahalalel and I saw in a vision how the sky collapsed-lowered, released-raised and fell to the earth. When it fell to the earth, I saw the earth swallowed up by a great abyss, [the] mountains suspended above [the] mountains, [the] hills lowered above [the] hills, and the great trees torn from their trunks, thrown and plunged into the abyss. That's why a [word] entered my mouth and I raised my voice to shout and said: "The earth is destroyed"! Then my grandfather Mahalalel woke me up, for I was lying near him; He said to me: "Why do you scream like that, my son, why do you utter such a lament [of fate]?" Then I told him all the vision I had had and he told me: "Just as you have seen a terrible thing, my son, since the vision of your dream about the mysteries of all the mysteries of the sins of the earth is terrible, so the earth is about to be devoured by the abysses and annihilated by a great destruction-ruin."* (1st Enoch 83:3-7) Such is the mystery that lies below, that the prophet Enoch wrote: *"The elect began to reside with the Chosen One and these are the measures that will be given for faith and that will strengthen justice." These measures will reveal all the mysteries of the depths of the earth and those who have been destroyed by the desert or swallowed by wild beasts or by the fish of the sea, those will be able to return supported by the day of the Chosen One, because none will be destroyed before the Lord Of the Spirits, none can be destroyed."* (1 Enoch 61:4-5) And likewise Ezra (Ezra) wrote: «*At his word the stars were fixed in their places, and he knows the number of the stars. He seeks the deep and its treasures, but has measured the sea and its contents; who has bounded the sea in the midst of the waters, and by his word <u>has suspended the earth above the water</u>. He has stretched out the sky like a dome and secured it over the waters ...*» (4th Ezra 16:56-59)

Adam wrote that even below there were places of light: «*[the] third hour [is that of] the worship of the deepest abysses, and of the light that is in the abysses, and of the light more hidden than man cannot penetrate."* (First Fragment of the Testament of Adam) It also seems to be an impassable place for men, or at least difficult to access, as the novels of yesteryear relate, when referring to the attempt of many heroes to go to Hades to rescue a loved one: «

They say to each other, reasoning wrongly: "Short and sad is our life, there is no remedy when man comes to an end, nor is anyone known to have returned from the Abyss." (Wisdom 2:1) And it also says: *"Because you have power over life and death, you bring down the gates of the Abyss and you bring up from there. Man, in his malice, can kill, but he does not bring back the spirit once it is gone, nor does it release the soul received by the Abyss."* (Wisdom 16:13-14) Job wrote: *"Men put an end to darkness, they examine everything perfectly, even the stones that are in darkness and in the shadow of death. They dig mines far from the inhabited, in forgotten places where no one he sets foot. There they are suspended, swaying away from the other men. From the earth comes bread, but within it is as if turned into fire, and in it there is a place where the stones are sapphire and the dust is gold. It is a path that no bird has ever seen, nor has the eye of a vulture seen it, that wild beasts have never trodden it, nor a lion has trodden it. Man puts his hand to the flint and overturns the mountains by their roots. In the crags he opens streams of waters, and His eyes see everything precious. He stops rivers at their source and brings to light what is hidden. "But where is wisdom found? Where is the place of intelligence? Man does not know its value, nor is it found in the land of living beings. The abyss says: "It is not in me", and the sea says: "It is not with me either". It will not be given in exchange for gold nor will its price be in weight of silver. It cannot be paid for with Ophir gold, precious onyx, or sapphire. [...] "Where, then, does wisdom come from and where is the place of intelligence? It is hidden from the eyes of all living things, and from every bird of the sky it is hidden! Abaddon and death say: "His fame has reached our ears."* (Job 28:3-22) In this part of Job 28, he seems to be talking about the rich and the VIOLENT (Job 27:13). That is, there are violent people, and this should be compared to one of the 8 parts in the Bible where "violent" appears, as beings that try to shake the things of God and creation: " *From the days of John the Baptist until now, the kingdom of heaven suffers violence, and the violent take it away."* (Matthew 11:12)

At some point, in a past time, God did things that are not fully explained in the Bible: « *At the breath of your breath the waters piled up; The currents gathered as in a heap; The abysses curdled in the middle of the sea."* (Exodus 15:8) Could that definition of "curdle" have something to do with "solidify"? In this sense we can also see: *"Then the depths of the waters appeared, And the foundations of the world were exposed, At your rebuke, O Lord, By the breath of your nostrils."* (Psalm 18:15) It is as if at some point he was seen down there -he or someone on his behalf- and everything that was in the midst of DARKNESS and DARKNESS was seen: «*He tilted the heavens, and descended; And there was thick darkness under his feet. He rode on a cherub, and flew; He flew on the wings of the wind."* (Psalm 18:9-10) And also in this regard we see that it seems as if he were going again recently or in the future: « *Behold, the heaven and the heaven of heavens, the abyss and the earth will be shaken at the hour of your visit."* (Sirach 16:18) We know that there is a lost Hebrew text about the deeds of Jehovah –a manuscript among many other missing ones-, called The Battles of Jehovah, which would possibly clarify all these events. Ergo, the ancients regularly personified, whether correctly or not, the elements of the cosmos. Seeing the parallelisms of Jehovah with the most important figures of mythology, we also find equivalences, which can mean that the events were also linked to the divine hand, although it did not directly act on it - or over time men erred, considering that their gods were the actors of this move. In other words, God acts through his Creation and nothing gets out of hand, and that is the reason why everything that happens is attributed, despite the fact that for a regular viewer there it does not seem that God was involved. The Most High allows things to take place because in the same way it is how he executes his designs and in the long run everything ends up directed to the end pre-established by the Creator from the beginning of the beginnings. This is how we see ourselves catapulted to another psalm of David: «*He who makes the winds his messengers, and the flames of fire his ministers. He founded the earth on its foundations; It will never be removed. With the abyss, as with a dress, you covered her; Over the Munts it was the waters. At your rebuke they fled; At the sound of your thunder they rushed; The mountains climbed, the valleys descended, to the place that you founded them. You have set an end for them, which they will not cross, nor will they cover the earth again.*" (Psalm 104:4-9)

By design of the Most High, the solar system would have been formed based on water, and under the masses of water it would have left holes that would be lined with hard rock: «For in his hand are *the depths of the earth, and the heights of the mountains are yours. His also the sea, for he made it; And his hands formed the dry land.*" (Psalm 95:4-5) This is what Enoch says: "*And then I made the heavenly circle firm, and caused the low waters which are under heaven to collect themselves together, into a hole, and that the chaos became dry, and it became so. Out of the waves (wave) I created hard and great rock, and from the rock I filled the dry, and the dry I called earth, and the middle of the earth I called the abyss, which is to say the bottomless (unfathomable), I collected the sea in a place and limited together with a yoke. And I said to the sea: Behold, I give you your eternal limits, and you must not break [,] let go of your component parts. So I quickly made the firmament. This day I called [for myself] the first-creation.*" (2nd Enoch 28:1-4) That "first creation" is the one that was destroyed with the Flood, according to some texts they show, leaving for us a new geological scenario from about 6,000 or 6,500 years ago. Another allusive passage is also from Enoch, and states: «*And this [is the] secret of this oath and [is] powerful their oath and on it [is] sustained the rakiyah (heaven/*

space), [because] for He always prophesied-dictated the world and even eternity. And the [planet] earth [has been] established on the waters and from [the] mysterious [of] the mountains come beautiful waters, from the creation of the world and to eternity. And that oath prophesied-ruled the sea and that the sand set a limit until [the] wrath is given and will not come from the creation of the world and until eternity. And in that strong powerful oath [are] [the] abysses, and they were standing still and did not move from their place, from eternity to eternity." (1 Enoch 69:16-19)

If that Abyss was established to separate what had appeared –that which did not come from the Most High-, it is necessary to know the forces that came accompanying this event, since evil was born there, which Wisdom itself (here called Faith), she strove to make amends: «... *when [Fe] saw [that] the shadow was stronger than her, she became jealous and, having conceived before herself, immediately generated Jealousy. This day came the beginning of jealousy in the aeons and worlds. This was [why] jealousy turns out to be a [characteristic that] lacks spirit. They had been born as shadows in a large amount of watery substance. [It was seen as a] bile coming out of the shadow, so [it] was then removed from the chaos. That day a watery substance arrived, and they were those who had penetrated it, [those] who appear in the chaos. As one who gives birth to a little one, [is] the one of all the surpluses of her fall and with the material that came from the shadows, she was expelled from her. The material was not of chaos, but it was in chaos: it is a part of it [that came to] be. After it came, then came Faith. She discovered the field of [all] chaos that had been thrown out as a [race] because it was meaningless, because it is, in fact, all boundless darkness and [the great] bottom of water. But when Fe [realized what] came from her disability, she was in trouble. And the disorder appeared as a formidable work. And he ran [into the chaos]. And she turned to him, [breathed] into his face, [which is] in the abyss in all the heavens."* (Apocryphon of Coptic

Creation 99:3 to 100:1) In ancient times everyone spoke of the darkness and darkness of the afterlife, of the great abyss and depth of the Earth, a gloomy and terrible place where malevolent spirits of the past dwell and where there is also the underworld for those who have left the body. Although the reference to "hell" is used to describe it, the usual word is the Greek Hades or the Hebrew Sheol. This idea grew the tendency to think that the dead man either went up to the divine abode, or was left to his fate in the afterlife. Despite the fact that this lurid and little-documented subject deserves its time to be studied, I will limit myself to mentioning the words of Job, in relation to the future of sinners: «Surely the light of the wicked will be extinguished, and the spark of their light will not shine. *fire.*» (Job 18:5) It does not say that they will be tormented, but it is a torment to be below. And this gentleman adds: " *From the light he will be thrown into the darkness, and cast out of the world.*" (Job 18:18) This may mean that the wicked will be removed from the surface of the Earth, either at death or at Judgment, as Enoch also pointed out.

Nature

We know the typical phrase that "Nature is wise", that "they are things of Nature" or about "Mother Nature", we could even find parallels with the Essene teachings of Jesus, when he spoke of the Earthly Mother, but the aspect it may not be the same even if it comes to be considered relevant. While Jesus referred to the Earthly Mother as the physical, harmonious part of Creation, and to the Heavenly Father as the spiritual, Shem's Paraphrase defines Nature as unconscious and manipulated by the forces of chaos. If Sem and Jesus were referring to the same thing, then we would have, once again, the personification of the Earth as a body or living being in itself, just as the ancient cultures understood it. Certainly the Earth is an enormous living organism, with its lungs in the Amazon, its heart in the Inner Earth, its blood in its bowels, like lava, its cells like each of the living species that walk on it, its oxygen and air like our own, its neurons like every human, and its nervous system like Ley Lines and geophysical energy. We would be able to cite a lot in this regard, but synthesizing we must point out that the similarity between all existence follows almost the same patterns, see from the prism that is seen: microorganisms, organisms of our size, macroorganisms, etc. At any type of scale, life continues, we observe it in our body, as an amount of small workers called atoms, molecules, cells, etc. and each one in its own dimension, one inside the other. At terrestrial and even stellar level it is the same. Everything follows a spectral and holographic sequence to whatever size it is seen, in the long run it is still life promoted by the vital spark of energy and spiritual power that emerges from the will of the Universal Father. For this and other conceptions Jesus taught: «*"The blood that flows in us is born from the blood of our Earthly Mother. Her blood falls from the clouds, springs from the bosom of the earth, murmurs in the mountain streams, flows spaciously in the rivers of the plains, it sleeps in the lakes and rages mighty in the stormy seas. "The air we breathe is born from the breath of our Earthly Mother. Its breath is sky blue in the heights of the heavens, it whistles on the mountain tops, it whispers among the leaves of the forest, it waves over the wheat fields, it dozes in the deep valleys and it burns in the desert. "The hardness of our bones is born from the bones of our Earthly Mother, from rocks and stones. They stand naked to the skies on top of the mountains, they are like giants lying asleep on the mountain slopes, as idols set up in the desert, and are hidden in the depths of the earth."The delicacy of our flesh is born of the flesh of our Earthly Mother; flesh that ripens yellow and red in the fruits of the trees, and feeds us in the furrows of the fields. "Our intestines are born from the intestines of our Earthly Mother, and are hidden from our eyes like the invisible depths of the earth. "The light of our eyes and the hearing of our ears are both born from the colors and sounds of our Earthly Mother, who envelops us like the waves of the sea around the fish, or like the swirling air around the bird."* (Gospel of the Essenes 1:19-28)

It is striking to observe that Jesus refers to material creation, nature, as a mother, a complement to the Father, not as his partner or as if he needed support, but as the material part of existing things. He personifies it as the active force, the existing and moving law and wisdom of what is created. In addition, he himself points out that death is the result of the violation of universal principles, damage to the virtues and qualities given by Mother Nature: «*And finally, the Son of Man will lose his life. For he did not keep the laws of his Mother, but added one sin to another. For this reason, all the gifts of the Earthly Mother are taken from her: breath, blood, bones, flesh, intestines, eyes and ears, and lastly, the life with which the Earthly Mother crowned her body.*» (Gospel of the Essenes 2:15-16) In addition to this, Christ reminds us that all created things obey spiritual laws and invisible forces, being also administered and coordinated by divine messengers and many entities that we do not know or understand. For this reason, the Lord spoke of the messengers that correspond to physical things, the elements, just as the prophet Enoch had described thousands of years before. Christ expressed it this way, orienting his teaching to the purity of the body and health: «*That is why you still cannot understand how much I speak to you about the Heavenly Father, who sent me to you. Follow then first only the laws of your Earthly Mother, of whom I have already told you. And when his angels have washed and renewed your bodies and strengthened your eyes, you will be able to bear the light of our Heavenly Father."* (Gospel of the Essenes 5:45-47) Jesus added, among many other words in this regard: «*"Truly I tell you that man is the son of the Earthly Mother, and from her the Son of Man received his whole body, from just as the newborn body is born from its mother's womb. Truly I tell you, you are one with the Earthly Mother; she is in you and you are in her. Of her you were born, in her you live, and to her you will return again. so much its laws, because no one can live long or be happy but he who honors his Earthly Mother and fulfills her laws.»* (Gospel of the Essenes 1:29:32)

Now, while Jesus taught the Essene sect of the Dead Sea about unity with what surrounds us, harmony, peace, purity, a healthy life and communion with the angels, Sem was exposed to the part related to the origins of Nature itself: «*And when the darkness saw [this] that (that is, the womb) she became a harlot, and when the water had been aroused, she rubbed her mother's womb. His mind dissolved into the depths of nature. It mixed with the power of the bitterness of the Darkness. And her (that is, the matrix of) eyes were broken in wickedness so that she could not bring back the mind. Because it was a seed of Nature of the dark root.*» (Paraphrase of Sem 4:25 to 5:1) The dark force that was in the Primal Waters, where Sophia's power had remained, was separated from itself, but the dark part ended up with the law of nature and the very power of her without Mother Earth, herself, perceiving this situation. All that potential, understood as the "Mind", which was the power of Sofia waiting to be restored, dispersed throughout the terrestrial creation, giving vital energy to many forms that came out of the powerful mind itself, since it was innate in its capacity, all time that it came from a superior and exalted being. Shem added: "*And when Nature had taken the mind for herself by means of dark energy, every image was formed in it. And when the darkness had acquired the image of the mind, it resembled the Spirit. But nature rose up to drive him out; [but] she was powerless against him, since she did not have a form of darkness. So [she] gave birth in the cloud. And the cloud shone.*" (Paraphrase of Shem 5:1-10) Consequently, many forms of life emerged, each having a special aspect, presumably visible in physicality. What they define over and over again as a "cloud" I would theorize is the result of all this space that contained this matter and energy, not being better identified with another word.

From that "cloud" would come the chaotic fire that the ancient peoples pointed to as the beginning of existence: *«A mind appeared in [it] like a terrible, damaging fire. (That is, the mind) collided with the unbegotten Spirit since they possess a likeness of it. In order that [nature] could become void of the chaotic fire and immediately after, nature was divided into 4 parts. They became the clouds that varied in their appearance. They were called hymen, placenta, energy, (and) water. And the hymen and the placenta and the power were the chaotic fires. And (that is, the mind) was inspired by the midst of darkness and water - since the mind was found in the midst of Nature and dark power - in order that noxious waters could not cling to it. Due to this [the] nature was divided, according to my will, in order that the mind may return to its power which [is] the dark root, which was mixed with it (i.e. the mind), had taken from him. And he (that is, the dark root) appeared in the womb. And in the division of Nature he separated from the dark energy that he possessed from the mind. (That is, the mind) entered through the middle of the energy - it was the middle region of Nature."* (Paraphrase of Sem 5:10 to 6:10) The need to separate the chaotic part of the waters was imperative, and the natural part of the primordial elements not combining shows why it was a priority to remove some things from others and situate certain "aspects" elsewhere, distancing them and placing them deep. The 4 "cloudy" divisions that became Nature are not without their representative importance, since the hymen is associated with virginity and the exit of what is gestated in the womb (that is why the uterus is also later spoken of). In turn, the placenta is the life support of a creature that is being gestated. None of this would sprout or flow by itself unless it had a driving force or energy that fueled it and an "embryonic" process. Those 3 elements are the ones that were described as "chaotic fire" while the water was the other part, which was then left only as an aquifer mass and whose volume we would define as oceans, either from the Earth or from other planetary "creatures", analogous. Apparently, the Mind, the power capable of creating and deciding, was in the middle of all that mixture of substances, energy and potential that is basically Nature, but that was also linked to darkness that is possibly a literal and symbolic state, that is, descriptive of the negative forces.

Later Shem writes: «*And Nature, which had been disturbed, immediately arose from the waters of inactivity. Since his promotion was a shame. And nature took for itself the power of fire. She became strong because of the light of the Spirit that was in nature. His image appeared in the water in the form of a hideous animal with many faces, which then twisted. A light came down into a chaos full of mist and dust, in order to do harm to nature. And the light of surprise found in the central region came to him after he got rid of the burden of Darkness. He was glad when the Spirit arose. Because I expected from the clouds down into the dark waters into the light that was in the depths of nature.*» (Shem 15:5-25 paraphrased) Repeating words already adopted in other manuscripts, the account focuses on what is unclear through mythology. This is advancing towards the aspect known to all of the separation of the water and the solid emergence of the mainland. It is the same referred to before about Lahmu (Lakhmu) and Lahamu (Lakhamu), two sister deities of the Mesopotamian vision, sons of Apsu and Tiamat. Normally known as the hairy or bearded, or also as the muddy. They are said to have had three pairs of locks and were naked except for a red triple girdle, and were often depicted as a serpent. It is also said that they represented the soil (mud), or the sediment, hence "the muddy ones", in addition to having been known as the thinkers. This makes clear the relevance of legends, even though they sound like "Chinese" tales, and how, personifying everything, they show deep knowledge and symbolism.

Referring to an identity of the deity, which manifested itself in various forms, one after the other, Shem wrote: «*Then, by the will of the Majesty, I took off the garment of light. I put on another formless garment of fire, which is of the mind of energy, which was separated, and which was prepared for me, according to my will, in the middle region. Well, the central region is covered with a dark power so that he could come and put it on. I went into chaos to save all light from her. Because without the power of darkness he could not oppose nature. But I leaned on his looking eye, which was a light of the Spirit. For this it had been prepared for me, as a garment and a rest for the Spirit. Through me, he opened his eyes to [the] Hades. Nature granted her voice for a while.*» (Paraphrase of Shem 18:1-25) When referring to his "garment" or "garment", whether it was light or fire, he seems to speak of a "body" or aspect with which he reveals himself, being only an "image" or "quality" corresponding to this dimension or to something capable of intervening in matter, since it is clearly spirit. It is curious that he talks about a garment that "*has no shape*", since technically the elements tend to adopt shapes in accordance with the environment. For example, water takes the form in which it is located, light the same, earth, air... and it is characteristic that it assumes a materialization with visible energy substances such as light or fire. That covering, disguise or casing was an idea developed in *the "mind of energy"*, the same one that had previously been separated. It is also essential to take into account that whoever tells this to Sem tells him that said being was expected to reveal himself and everything was determined for him. Also delimiting that said matter was his will, although he does not identify himself as God -since he speaks of "*the Majesty*" as the one who directs destiny and determinations, and places it outside of Him as another entity- because it also refers to Him in the third person. Then, when speaking of what is «*in the middle region*», it opens the way to the question of what this "middle region" is. If it is "middle" then it is in the middle, between two places, therefore it is necessary to consider the aspect of the abyss as the lowest, and the celestial as the highest, leaving the Earth in the middle -since the same waters are often identified, namely "down" and "up."

The translation notes that "*the central region is covered with a dark power.*" Will the "central region" be the same as the "middle region"? An admissible question in translation or as a synonym. That area is the one that is "covered" with a "power" of "dark" essence. According to the thread of the recitation, this place must correspond with Nature, the terrestrial surface and the life interconnected in it and that vibrates in itself, although it is subject to that power or negative energy. In the next part, that divine being who covers himself with light and fire, says that he has descended into " *chaos to save all the light from it.*" This story is reminiscent of Sophia's descent, but here either it connects it with Christ or someone close to him from the great imperishable powers, or it refers to Sophia as united to said entity, or it was simply a subsequent event, which would have taken place after let yourself see the light of Sofia. Who narrates said facts to Sem clarifies that removing power from chaos was how Nature could defend itself and get rid of the adjacent and incorporated darkness. It is as if death-like laws had been established and could not be broken until a divine figure came down to wrest from said dark force the power to subject all to death. The divine entity mentioned said to have "leaned" on the observing eye, in front of it, whom he refers to as "*a light of the Spirit.*" Was he referring to Sofía, implying that she sees everything? Another detail within this predestined event was that the entity, which seems to correspond to Christ, confesses that it had also been given a "garment" and "*a rest by the Spirit.*" The word "rest", in Hebrew, can be associated with "noah", "janoch" or "shabbat". He also adds that thanks to this being, that light was able to see clearly towards Hades, the lower parts of the Earth and " *nature granted her voice for a time"*, as if she wanted to imply that the Law of Nature and your motor Mind would have "the word" for a certain age, period or time, ruling according to its own innate parameters.

«*And my garment of fire, according to the will of the Majesty, went to what is strong, and to the dirty part of Nature, which (the) power of darkness covered. And my clothes rubbed [with] Nature on its cover. And her filthy femininity was strong. And the belly of anger rose and made a dry count, similar to a fish that has a drop of fire and a fire power. And when Nature had cast him out of mind, he was troubled and wept. When she was wounded, and in tears, she cast off the power of the Holy Spirit (and) kept as I put myself in the light of the Spirit and rested with my clothes because of the eyes of the fish.*» (Paraphrase of Sem 18:25 to 19:10) It is honestly difficult to understand what is being spoken, since that archaic vocabulary limits the possibilities of understanding, which today, with a wider range of words, could possibly be better expressed manner. In any case, the person speaking expresses that the "body" that he possesses, or possessed, had some type of "contact" with Nature which is described as not light but active, specifically with the part that proliferated in force. That area of Nature was the one that was black, and this section was indicated as "feminine", but in relation to an unclean tendency and a characteristic polarity in its behavior. Now then, there is talk of a "belly" again, defined with the quality of being "anger", which "rises", possibly from the lower parts, reflecting the "dry" aspect. This belly, in its action, is compared to a giant fish with two characteristics or elemental "sparks": a drop of fire and a power of fire. What is a drop of fire? Keeping in mind that the word that the translator took may have meant not only "drop" but "fall", let us suppose that it speaks of a potential in two appearances. That event identifies or reflects the differentiating action that would free Nature from that dark yoke that is at the same time uprooted from the Mind, although... why did he cry? The wound that is expressed, would it be emotional or physical? In any case, it seemed as if in this way the potential of the Holy Spirit was expelled from those essences, which would come to be next to the light. And about the last thing, what do you mean by « *the eyes of the fish*»? If the creation of the fish had not yet been discussed, what fish is he talking about and why should he rest from their sight?

«*And since the works of nature could be condemned, since she is blind, multiple animals came out of her, according to the number of fleeting winds. All of them came into existence in Hades in search of the light of the mind that was taking shape. They weren't able to stand up against her. I gloated over your ignorance. I will find myself, the son of the Majesty, in front of the womb that has many forms. I stood [in front of] the beast, and placed before it a great demand that heaven and earth might become, in order that all light might arise. For in no other way could the power of the Holy Spirit save one from slavery, except by appearing to it in animal form. Therefore it was compassion for me as if I were his son.*» (Paraphrase of Sem 19:10-35) Now we begin to talk about condemnation, firstly about the actions of Nature. They wanted to avoid being singled out or judged, since Nature seems to lack conscience and has been subdued by darkness against its will. Therefore, the emergence and proliferation of creatures that arose from Nature, of a multitude of varieties, took place. However, his specific origin was not the surface but the lower parts of the Earth, and his interest was behind " *the light of the mind* " that was beginning to be defined. That is to say, the obedience of these forms of life was subject to the parameter that was being developed at that moment. What is not very logical is why they make fun of their ignorance; These "individuals" have all the earmarks of being "dark entities" of some kind (I think those known in ancient times as "spirits of the water").

It is strange to think of animal species that are in the underworld and that also act in this way. That is why I believe that it refers to souls, a species of creatures that folklore defines as a type of "demons" or simply "spirits". What I propose is the idea that is ventilated in Taoism and animism, where it is believed in the existence of spirits of Nature. On the other hand, the womb is described as that « *which has many forms* », just like how Sakla is represented, and apart from that the "Son" puts before the Beast the complaint about how the existent has been created, making it evident that the The only way the Holy Spirit was freed from slavery was through animal forms, which once again brings us back to the representations applied in the Law of Moses and the sacrifices.

Assuming that the one who reveals this to Shem is, in some way, Christ (accepting that he had not yet been born in the flesh) or a messenger from him, the version of this terrestrial creation argues further, saying: «And because of my request, *nature arose from possessing the power of the Spirit of night and fire. Since he had removed his forms. When she had left him, he blew on the water. Heaven has been created. And from the foam of heaven the earth came into being. And in my desire they brought forth all kinds of food, according to the number of the beasts. And she brought forth the dew of the winds on account of you and those who will be born a second time on the earth. Because the earth possessed chaotic fire power. Thus, he gave birth to every seed."* (Paraphrase of Sem 20:1-20) The decision seems empirically proper to the Son, although having a spiritual essence of two qualities: night and fire, since fire can also be understood as "energy" in a prehistoric vocabulary. By saying *"that the forms had been removed"*, he seems to emphasize his initial or autochthonous composition, moving on to a different substance or image. Once this was done, it was immediately blown over the existing water and the sky was created, that is, the separation of the water brought these elements upwards, leaving a space or void in between that cloisters what we understand by "sky". Next, and consequently, he speaks *of "the foam of heaven"* as the foundation or principle from which *"the earth came into being."* Remembering the Greek legend that narrates the castration that Cronus performs on his father Urano, we see that the mutilated genitals of Heaven (Uranus) fall into the sea, and Aphrodite (foam deity) emerges on the shore. The parallelism is significant and this story is also repeated in Mesopotamian mythology and, with a certain similarity, in Hindu mythology, when Indra, defeating the dragon, claims his father's place and dethrones him. Then, in the Gnostic texts, they say that one of the 7 of Sakla, specifically Sabaot, Lord of Armies -the same name as Indra- rebels against his father and unleashes war in heaven against him and his brothers, being helped by Sofia and subjecting everyone to her power. Now, the genitals symbolize the motor of the progeny, while the uterus, the womb, the ovaries and the ovum are emblems of the gestation, formation and preparation of something, technically proper to the emerging "life". In another order, the fact that the Earth took place thanks to the "foam" of Heaven is a fact that makes one imagine an atmosphere being born full of agglomerated substances, many more per square meter than what pure air currently has. There are those who have suggested that the Earth's atmosphere or sky in the era before the Flood, and millennia before, was increasingly solid, that is, composed of many more elements than are observable today the more one travels retrograde. A global event of water boiling or breaking up to make way for a lower dry part could also be visualized, resulting in a formidable spectacle similar to a titanic cloud of foam, dew, moisture, etc.

An event of this magnitude resembles a planetary explosion. The Russian Sumerologist Zecharia Sitchin, an expert in cuneiform writing, stated in several of his 13 books that the mythical battles referred to in the Babylonian, Assyrian, Akkadian and Sumerian novels about origins, in which the primordial gods fought, were nothing more than the impact of stellar bodies, one against the other, in the incursion of new objects into the system in which we are. This neighborhood would be, at that time, in full birth, but said impacts would have burst celestial bodies and in turn formed new floating masses, guided by the gravitational force of the rest. This theory is plausible to a certain degree, although it lacks a few points that do not appear in the other accounts, such as the fact that the others do not speak of alien masses invading the watery orb from which Heaven and Earth arose—or the less so it seems. Thus, with this dynamic of accidents came the appearance of plant life, aimed at feeding those who were to come to live in this new world, following a calculation defined in the proportion of the variety: the same number as the animals (a clear example of balance in biodiversity). So the dew appears, surely to promote life in the plants and trees destined for consumption, which scientifically could be the secondary effect of the masses or exorbitant amounts of water moving away from the initial point and rising over the heights of the Earth. It would not be strange then that many myths identified the primeval world as a vast amount of humid and muddy territories, with an infinite number of seas, lakes, rivers, waterfalls, springs, mangroves, galleries with water tanks, and many other similar ones. Sum that was manifested thanks to the *"chaotic fire power"* that this orb possessed, resulting in the indicated way all kinds of producing seeds with a pre-established and predetermined purpose.

«And when heaven and earth were created, my garment of fire arose in the midst of the cloud of Nature (and) shone over the whole world until nature became dry. The darkness that was his garment (that is, of the earth), was thrown into the perilous waters. The central region was cleansed of Darkness. But the belly [was] annoyed with

what had happened. It is perceived in its parts what was the water as a mirror. When he perceives (him), he wonders how he had come to be. Therefore she was left a widow. He was also surprised (that) he was not in it. For the forms still possessed a power of fire and light. (That is, the power) remained in order that it might be in nature until all powers were separated from it. Because just as the light of the Spirit was completed in three clouds, it is also necessary that the power that is in Hades will be completed at the appointed time. For, by the grace of the Majesty, I have come to her in the water for the second time. Well, my face satisfied with her. His face was happy too." (Paraphrase of Sem 20:20 to 21:15) Once Heaven and Earth were created, this spirit, taking a hot substance as its physical expression, appeared in that "cloud" of Nature, illuminating the aquifer world until it came to reveal the dry parts. Then the bad, negative part was expelled from the Earth, being destined for *"dangerous waters"*, although what are those dangerous waters? The nether bottoms? It seems that this darkness was the bark of the dry part that the water covered or, in another sense, the darkness was an essence that wanted to be suppressed from this creation and, therefore, was removed from the elements, leaving in its place a solid mass, as if the moisture were sucked out of the mud, leaving only aridity. This is how that so-called "central region" got rid of the dark part, although this seemed to have hurt the "belly". Now it is perceived that the water, before this, had other characteristics or substance, but once this fact took place, this liquid element began to be seen as crystal (diaphanous), so it is presumed that before it was gloomy or gloomy. At this point we observe that power, Nature and/or mind did not know what was happening, how it came to be what it was and/or how the darkness was taken from it. By saying that *"she was left a widow"*, she reminds us of Tiamat (the Earth) having lost Apsu (the primordial ocean) -in Mesopotamian mythology-, who was supposed to be her consort, that is, her companion or complement. The fact that this terrestrial essence or entity was reflected thinking or discerning makes clear what was written: "*it still possessed a power of fire and light.*" Although this can identify, as I have already mentioned, pure energy, it also transcends beyond, attributing itself to being a virtue or inherence as a spiritual creature, which should not be lost until everything returns to normal.

After the mention of the "*three clouds*", I, as an idea, would suggest that it is the next step of the dismantling, letting itself be seen in 3 masses from which other planets would continue to form. All this happened "in front of the waters", acting on them, in the middle of them and under them, substituting the elements and forces contained in them. Therefore, the next step had to be revealed, showing an objective well considered of old: «*And I said to him, "the power of seed in the earth must come from you." And she obeyed the will of the Spirit that could be reduced to nothing. And when they returned their ways, they rubbed their tongues against each other; they copulated; they engendered the spirits and demons and the power that is of fire and darkness and the Spirit. But the form that was left alone cast the [of] the beast itself. She wasn't having sex, but she was the one who rubbed herself. And she gave birth to a spirit that possessed a power of fire and darkness and the Spirit.*" (Paraphrase of Sem 21:15-35) The order was clear: "provoke life" and that life had the function of "feeding", but who? It had to be installed for those who had previously been arranged to come. So this was all deliberate. The " *seed power* " is a kind of spirit in Nature that has the ability to motivate the bark to generate grains, pipes, seeds... a whole code within a tiny receptacle that, once a tree has been produced, its fruit put this tiny package back on the ground so that it can produce again. A whole genius... But well, what does he mean by " *he rubbed his tongue against each other* "? Not only is a movement motivated in the "language", but it does so among everything that is in it, emphasizing this process as a "copulation". Said action resulted in the appearance of spirits, demons and powers, which are born of this force and typical of 3 primordial substances: fire, darkness and spirit. All these brushed against each other, but the only one that did not mate with another form was the Beast. Believing he was different, he rubbed himself, something like masturbation, a not unusual aspect because this is how the myth of Ra is expressed in Egyptian culture, saying that in this way the main god, this solar being, produced the life of other people. deities. Specifically, Sakla, who is the Beast, brought into existence, thanks to this unaccompanied action, a demon with the same characteristics and essence as those others who were with him at the beginning, although who he is is not defined.

Such a barbarity and abomination did not seem to go to "good port", foreseeing something catastrophic for creation: «*And so that the demons could also become free from the power they possessed through impure sexual intercourse, a womb was with the spirits of the water it resembles. And a dirty penis was with the demons, according to the example of darkness, and in the form in which it is rubbed with the womb from the very beginning. And after the forms of Nature had been together, they separated from each other. So they threw her out of power, marveling at the deception that had befallen them. They were afflicted with an eternal pain. They covered themselves with his power.*" (Paraphrase of Sem 22:1-15) It seems that the way in which the power that had also passed -synthesized and diluted more and more- to the creations ended up being removed to the dark, it was determined that the impure sex between these Forces would be the way, thus managing to take power away from them definitively. For this reason, the representation of the penis and the action of disordered sex between the spirits and demons are made, which, once rubbed against each other, would have followed in the footsteps of Sakla, "rubbing" themselves, alone. They tried to imitate the things that were happening. Something in them gave them a glimpse of insight into the pattern by which they and the things around them were created, but they were still ignorant of deity and higher powers. It seems that once expelled from the essence of Nature, they were arrogant because they saw themselves as strong, but realizing what was really happening, suppressed from the essence of Nature, they became ashamed and had nothing more to hide behind their own justification. using his qualities to hide his full-fledged nudity.

«And when I had put them to shame, I got up with my clothes in power and - what is above the beast that is a light, so that I could make the desolate nature. The mind that had appeared in the nature of darkness, (and) that was the eye of the heart of darkness, in my desire to reign in spirits and demons. And gave him an image of fire, light, and attention, and a part of naive word. Therefore he was given the greatness of [itself] in order to be strong in his power, independent of the power, independent of the light of the Spirit, and relations of darkness, in order that, in the end of times, when nature will be destroyed, he can rest in the place of honor. Because he is found faithful, since he has hated nature's unchastity with darkness. The strong power of the mind arose from the mind and the spirit begotten.» (Paraphrase of Shem 22:15 to 23:5) Whoever refers this story to Shem affirms that when this took place the next thing was to take his " *clothes IN POWER* ", as if he returned to his true supreme essence or as if he directly took control over everything. With this he intended to be a light superior to the Beast and devastate Nature. What would he mean by " *could make Nature desolate* "? That identifies an abandonment, loneliness or emptiness. It seems that the Mind that had been at the heart of this creation must be free of darkness and be the one who reigns over spirits and demons. He defines this circumstance as the " *eye of the heart* ", as a nerve center, heart or focus of direction, while the eye represents

observance and vigilance, and the heart stimuli and feelings. Apparently, this guiding Mind would be accompanied by virtues, such as an *"image of fire, light"* and *"naive word"*, assuming that innocence would free the Mind that directs Nature from judgment.

This story is summarized in a text attributed to Asclepius (said to be the son of a mortal named Coronis or Coronide, who became pregnant by the solar god Apollo), from a revelation given by Hermes Trismegistus, in which he says: "There *was for in the abyss an immeasurable Darkness, and a water and a subtle intelligent spirit: the divine power kept them in Chaos. Then a pure Light emerged that condensed the elements under the sand, extracting them from the humid substance,... and all the gods separated from nature full of seeds. When all things were indefinite and unformed, the light ones separated upwards, the heavy ones rested on the bottom of humid sand, and by the action of the fire each and every one of the things were defined, and they remained suspended in order to let the spirit guide them. Heaven was seen in 7 circles, and the gods were shown in the form of stars with all their constellations, and... (the structure?)... was organized with the gods that were in it; and the orb, at its periphery, turned round in the air, driven in its circular course by the divine spirit.*" (Hermes Trismegistus to Asclepius, Corpus Hermeticum. Treatise XVII, verses 1-2. Original incomplete and without title) In a different text, but with the same title, Asclepius wrote: «And having said these things, he changed form, and *in For an instant the entire space opened before me, and I saw an infinite panorama, and everything was transformed into Light, a Light so serene and joyful that when I saw it I adored it. After a short time, a frightening and gloomy Darkness began to descend, coiled like a tortuous spiral, similar to a serpent. Afterwards, the Darkness gradually transformed into a certain humid nature that stirred indescribably, that spewed smoke as fire does and emitted a cry, an inexpressible moan. From there came an inarticulate cry for help that sounded like the voice of a human being. It was then when, coming out of the Light, a holy Name fell on the thing, and a pure fire emerged from that humid nature towards the celestial spaces, a light and subtle fire, and energetic at the same time. The nimble air was carried away by the spirit, and from the earth and the water it lifted itself up to the fire, so that it seemed to hang from it. For their part, the earth and the water were so closely intermingled that it was not possible to distinguish one from the other: the spiritual Name that hovered over them kept them in motion, to what seemed to be heard.»*

« Then Poimandres told me:- Do you understand what this vision means? - I will know, I answered. - I am that Light, he told me, I, the Mind, your God, who pre-existed the humid nature that emerged from Darkness. On the other hand, the luminous Name that proceeds from the Mind is the son of God. - And then?, I exclaimed. - Understand it like this: what you see and hear is the name of the Lord, your mind instead is God the Father, since they are not mutually separated, since their unity is Life. [...] I was still astonished, when he spoke to me again and said: - Have you mentally seen the archetypal form, the principle prior to the unlimited principle, this was what Poimandres told me and I asked him: - Where did the elements of the nature? And he in turn:- Of the Will of God who, having welcomed the Name and contemplated the beautiful cosmos, imitated it, creating for himself from his own elements and from the souls made by her. The Mind the God, which is both male and female, and contains in itself Light and Life, gave birth by Name to a second Creative Mind, which, being god of fire and spirit, created in turn seven governors container owners of the sensitive cosmos, whose government is called Destiny. Immediately, the Name of the God, tearing itself from the lower elements of the God, launched itself towards the pure region of created nature and united itself with the creative Mind (since they are of the same nature), leaving the lower elements of the created nature helpless. nature, the irrational ones, which consist of only matter. Then

the Creative Mind, together with the Name, wrapped the circles and made them spin roaring, they set their own creatures in circular motion so that they would roll, from an indefinite beginning, to an endless term, which begins where it ends. This circulation of everything, as the Mind wanted it, produced irrational animals from inferior elements (the Name was no longer with them), the air produced birds and the water fish. The earth and the water, as the Mind wanted, were separated from each other, and the earth made the animals that it had inside come out of itself, quadrupeds and reptiles, wild beasts and domestic animals.» (Poimanides to Asclepius, Corpus Hermeticum. Treatise XVII, verses 4-11. Original incomplete and without title)

In another writing, where it is Hermes who instructs Asclepius - as well as King Amun and his own son (of Hermes) - it was mentioned: «*It was once God and "Hylé", a Greek word that we translate as Matter. The Spirit was next to Matter, or rather was within Matter, although not in the same way as it was in God nor as those (principles?, gods?, essences?) from which the world was born were in God . Because they were not yet born, although they already existed in Him, from where they would later be born. And by those who "were not born", we do not want to refer only to those who had not yet come to be born, but also to those who lack the ability to generate, that is, from whom nothing can be born. Because all those who possess the power to engender are generators and one can be born from them, even though they were not born from themselves. (Because no one doubts that one can easily be born from those who were born from themselves and from whom all things are born). Consequently, the everlasting God cannot and could not be generated: so it was, so it is, and so it will always be. This is the nature of God, an entire proceeding of himself. On the other hand, "Hylé", the nature of Matter, and the Spirit, although they do not appear as engendered from a principle, nevertheless possess the capacity, in themselves, to be born and to engender. The beginning of fertility is in the way of being of their natures, which possess in themselves the strength and reality of conceiving and giving birth. Where alone, then, they are capable of generating, without the help of anyone who makes them conceive. On the other hand, regarding the things that cannot be conceived without the help of the union with another, it is necessary to think of them in such a way that we consider the space or place of the world and all the things it contains as unbegotten, because in itself it contains the universal power. of generation. I speak of the space in which all things are, because nothing could have existed without the space that could contain them all - nothing could exist if it had not been assigned a place before -: one could not speak of qualities, nor of sizes, nor of positions, nor of activities of things that are nowhere.»*

Therefore, Matter, although not engendered, contains in itself the nature of all things and provides all of them with an inexhaustibly fecund matrix. This is the entire virtue of uncreated Matter: the power to create. But nevertheless, just as its nature is fruitful, so it is equally fruitful in evil. Did I not say to you, oh Asclepius and Amon, what many repeat: "God could not have abolished evil and removed it from the nature of things"? But there is nothing that can be answered. But for you I will continue what I have begun, and I will explain it to you. So they say that God should have rid the world of any kind of evil, which nevertheless is installed in the world as his member. However God provided and took care, as much as possible, deigning to grant man mind, science and reason. And because of these things, which grant us pre-eminence over other living beings, we are the only ones able to avoid the tricks, traps and vices of Matter, and, to avoid them as soon as they want to appear, intelligence and intelligence were granted to man. prudence, because the foundation of all science resides in the utmost Goodness. All the things that are in the world are governed and live by the Spirit, which behaves like an organ or instrument submitted to the supreme will of God. It is sufficient up to here what has been discussed." (From Hermes Trismegistus addressed to Asclepius, verses 14-16. Corpus Hermeticum)

Eros

We could find similarities of the idea of Aphrodite, of the idea of Lilith, and of the idea of Eros and more archaic legends in line with the revelation that Shem, son of Noah, had. From the initial chaos that came out of the Mind that the darkness subdued, it appreciated a cloud in whose interior a matter was being forged, « *And when the darkness saw this -that is, the womb- it became a harlot, and when the water had awakened, he rubbed his mother's womb. His mind dissolved into the depths of nature. It mixed with the power of the bitterness of the Darkness. And the eyes of her (that is, the womb) were broken in wickedness in order that she could not bring her mind back. Because it was a seed of Nature of the dark root.* » (Paraphrase of Sem 4:25 to 5:1) The situation of the harlot is abominable before the healthy sexual bond. The point of this activity is that it is considered corrupt because the body is rented and the harmony of the unique union is sullied. By defining darkness as a harlot, it is as if it were identified with those who give themselves to others, through fornication, without caring about their own integrity, since their objective, more in this case, is profit. Sakla showed here to be so blinded in his stubbornness and to be so obtuse that he was only interested in his preeminent and arrogant place always existing, no matter what. Then, according to what seems to be understood from this writing, when it was the right moment, when the "water" emerged from all this, it was the hour in which that great womb dissolved and manifested the matter over which the Mind itself of power it would melt away, becoming one with Nature. That event would be the beginning of the existence of what some have called Mother Earth or Mother Nature, truly blinded by darkness – both intertwined out of sight of the human eye. Now, what follows next is the part of copulation and the union of polarities, for which a motor force came into existence, in the form of an entity or spirit, which would be in charge of carrying this out at all levels! and in all life systems, whether physical, astral or mental. That creature is known as Eros, which in Greek identifies one of the 3 existing types of love, specifically passionate.

Among the aspects that took place at the beginning of the universe, passion began to act, also being one of many existing forms that arose, along with Death, Fate, Jealousy, Envy and Anger. Defined in Greek as Eros, passion has been the catalyst for the main evils of this globe, motivating all kinds of sexual imbalances, obsessions and perversions, even though it had an elementary principle aimed at complementing the species in their opposites. Solomon spoke of this being in a personified way: «*And when I answered him that I [will not give] any worship [to] strange gods, he told the maiden not to sleep with me until I fulfilled and sacrificed to the gods. Then he went to the cunning Eros and brought for me his 5 grasshoppers, saying: "Take these grasshoppers, and crush them together in the name of the god Molech, and then I will sleep with you".*» (Testament of Solomon 1:129) This lady that Solomon wanted to court was a foreigner and motivated him to build temples to strange gods, an aspect that began the decline of the monarch and his kingdom. It is notorious that she uses Eros, passion, to captivate Solomon and persuade him. So where does Eros come from? In Greek mythology there are at least 3 Eros recorded: One is the son of the goddess Aphrodite and Hermes (although others say that he was with Hephaestus/Vulcan), and he is popularly known as Cupid (it is assumed that he is the father of the famous Hermes Trismegistus, credited with founding Hermetics), brother of Hermaphroditus –although others assume he is the same. The other Eros would be the son of Heaven (Uranus), according to one version; son of Nyx (Night) and Erebo, according to another; and offspring of Night and Chaos on the other (the genealogies of Greek mythology are crazy, more jumbled and uncertain than the Egyptian ones). The version that says that Eros is a descendant of Night and Chaos attributes the birth of the Earth to him. This diction is consistent with the story by which we are on our way.

Well, we find here many meanings to urban legends about love, romance, sex, falling in love, lust and sensuality. However, the first Eros to enter the scene is a personified force, allegedly the son of Erebus, who was the very personification of Darkness. Let's remember that this Erebus was sent to the underworld where he became the Infernal Shadows, apparently for having helped the titans in the war against the gods. It is curious to note the parallelism of the Greek myth with the story of Sakla, since Erebo's consort was Nyx, who is precisely Night. The ancient Hellenes would have written that from these arose the High Heaven and the Day, which we would know to consider as a late misrepresentation of the true facts (where Moses refers to God making the Heavens and it was the First Day), but chronologically similar to the Biblical version of High Heaven (Shamaim), distinct from atmospheric heaven, and Day, as "Yom" first. The striking issue here is that the Hellenes would have recorded that supposedly Eros (Passion) joined Nyx (Night) and from this Gea arose, also called Gaia (Earth), the mother of the titans (giants). Eros can represent all the emotional, biological, genetic, social and psychological cataclysm that stirred and was born with concupiscences, lasciviousness, copulation and other sexual elements, even before being manifested in the flesh. It can also symbolize, in a generic or globalized way, the fall of the children of heaven and their union and copulation with humans. We have read that the light that Moses spoke of in his first book was the manifestation of the Ogdoad that came from the upper part of the existing veil between realities, shining in the chaos and darkness. This apparition exposed Sakla and put the gods in trouble, but it would not be the first nor the only intervention to put an end to them. The Egyptian text continues affirming that it was determined to let an angelic creature appear that would reconcile the negative elements with the existing creation: «*As soon as Providence saw that angel, she fell in love with him, because she hated being in the dark. And I wanted to hug her, but she didn't get to [get it]. [He was] unable to curb his love, and his light has spread on earth. From the days, this angel was called "Adam-Light", which means "bright-blooded-man", and the spread on the ground is "holy Adamah", which means "holy ground adamantine". From that date, the authorities feared the blood of the virgin. But it became the pure land, because [they were] of white blood. And more of the water was pure grace to reflect Faith-Wisdom, which appeared to the Great Bull in the water. Therefore, it is fair to say "by the waters", of water that is holy, since it animates all that is purified.*" (Rebellion Apocryphon 108:15-33)

The gods feared the lineage of light, considering that if it materialized it would reduce them. Providence or Fate tried to catch the brilliance coming from superior realities, but could not. Sakla did see it more closely, but nevertheless, it was through the reflection in the water. Since then it was known that there was a superior lineage that the gods feared would be manifest, an unsullied, virgin lineage, which was the perfect blood from above, defined as Adam, the genetic code of light. Although Sofia appeared, and it even seems that Christ did too - all of them in the "name" of the Father -, also this new creature that enters the scene, which comes to be in charge of the union of the polarities in Nature. Now, something strange seems to take place at that moment: «*From that first blood appeared Eros androgynous. His masculinity, Himéros, is from the fire of light. [The] femininity that accompanies it is a number blood soul of the substance of Providence.*

He is so enchanting in beauty, more graceful than all the creatures of chaos. Once they saw Eros, the gods and their angels became lovers of him. And, when it appeared among them, [it was] in flames. From a lamp to several lights, and although this point of view is unique, the lamp does not [produce] blindness... on this path also Eros [was] divided among all the creatures of chaos and continues without ceasing. In the same way only [he is] the intermediary space between light and darkness, Eros appeared, through the angels and men it was that the union of Eros was achieved - in the same way that on earth the first germinate delight. The woman followed the land, and the marriage followed the woman, the generation of the marriage, dissolution followed by generation.» (Apocryphon of the Rebellion 109:1-25) Ergo, apparently something happened at that moment and from said divine blood that came to reproduce in the now consolidated ground (which is why it was defined as Adamá) came Eros, who would be an intermediary between light and darkness, between higher and lower realities, an aspect which is intrinsically related to sexual matters or the union between feminine and masculine concepts, elements or entities. It is not by chance that the masculine part of Eros was Hímeros, and this brings us to Greek literature, where indeed Hemera, the Day, was Eros' half-brother. It must be seen that it was already understood in Hellenic culture that Eros represented beauty and passion, which is why Aphrodite herself named her most beautiful son that way. Procreation thanks to passion seems to have started in this way and creation itself came to follow this pattern, basically by the union of a masculine and a feminine element, especially thanks to sexual passion – but bringing the hidden "Mystery of Marriage", to sanctify.

Another interesting matter is the fact that Aphrodite (Astarte), according to a myth from the Peloponnese and Macedonia, would have come from Asia Minor and joined Olympus. Another version, the most common, said that Cronos castrated his father Urano and threw his genitals into the sea. This tale states that Urano's genitals, reaching the seashore, turned into a foam from which Aphrodite emerged (that is why she received that name: "afro" is "foam", and "diti" comes from the Indo-European: "deity" or "divinity"). Although both accounts suggest that Zeus, admired by the beauty of Aphrodite, incorporated her into the pantheon as one of the 12 Olympian gods. Although she was married to the deformed Vulcan (Hephaistos), she had countless adventures with gods and even with demi-gods (understood as heroes). It is also important to consider the fact that Aphrodite seems to change from an aggressive and warlike way to an entirely vain and materialistic one, oriented towards physical beauty. For this reason, she is associated with the Canaanite warrior Astarte and, as if she had tired of this canon, she left that region and assumed a passive role (hence, in Egypt they see her as the assassin Sakhmet or Sekhmet who became a goddess of love, renouncing conflicts and blood, then being known as Hathor). On the other hand, and returning to Eros, it is inspiring that they say of Eros that "*it is] the intermediary space between light and darkness*", assuming that it has some resemblance to the "rakiyah" (space) that differentiated light of darkness, since this implies the physical separation –and perhaps of non-physical elements- between waters and waters. One more curiosity, in reference to the previous sentence, is the question of his saying that " *through the angels and men the union of Eros was achieved* ", showing the parallelism with the earth, where " *the former germinate delight.*» It is rash to suppose that he said that Eros was "completed" in his work, achieving his goal thanks to angels and humans, but in what sense? And what relation would it have to generate "desire" or "delight"? The answer may lie in the next part, which states: "*The woman followed the land, and the marriage followed the woman, the generation of marriage, dissolution followed by generation."*» This subsequent detail refers to the fact that the feminine apparition came after the Earth with an important complementary function, behind which the "union" (marriage) also walked, initiating something that seemed not to exist before.

This they reflect as a "dissolution" which is followed by "generation", that is, gender, progeny, development, etc. We can assume that there is talk of the bond between man and woman, the sexual aspect (not as copulation without commitment, but legitimate and official), so that behind this there must be a mystery that implies the relevance of using this bond to suppress the deficiency, by means of an aspect that does not exist -in this way- in the superior realities, and that, glimpsing its relevance, has been used by negative forces to destroy its true purpose and degenerate it. All of this is symbolically very similar to the Egyptian account, so there may be some kind of root or reminiscence.

The text continues announcing about Eros: «*After [arise] this Eros was that the vine grew from the blood that was spilled on the ground. That is why people who drink conceive the desire to... After the vine, the fig and Grenadier germinated in the earth with the rest of the trees according to their kind, thus in their seeds, after the seeds of the authorities and their angels."* (Apocryphon of the Rebellion 109:26 to 110:1) When it says that *"they conceive the desire of ...",* there is a possible word that goes there, according to the chinks of the translation, such as "kill" or "dull". This seems to explain the part of the emergence of plants and seeds on Earth, which Moses referred to in the Torah. The vegetation would appear as the fruit of the seed of the gods and Eros on the ground, something that expressed his own feelings. The manuscript points out that after Eros appeared, he does his job, resulting in " *the vine* " growing " *from the blood* ", as if Eros had decomposed or defragmented, mixing with Mother Earth through the forces and powers intertwined with the items. We could say it in another way: a natural law is created. The intrusion of Eros generates life in these canons, understood as the motor that produces thanks to the union of the masculine and feminine elements. It is notorious, and patent, in Nature, in animals, plants, even in the winds and clouds –for those who know about these matters. It is like the hot and cold currents that exist, the opposites that together create and bear fruit, just like the oriental idea of Yin Yang (male and female – heaven and earth) and the duality necessary for universal balance. Furthermore, it is not strange that the Earth, since it was known as Aretz, began to be defined as Adamá, either because it was the abode of the Adamics and/or because of what the definition itself means: "divine bloody", "adamic", "[land of] Adam", "the one of divine blood", and a few other epithets that can come from there. What follows reflects the natural law that every inhabitant of the world is attracted to that "terrestrial" rule of uniting between opposite poles: male and female. The clear example of this, expressed by the great power in Eros, produced life from the Earth itself, under the parameters of this generating spirit and in charge of consolidating the polarities and uniting them, causing, in the long term, the dark to begin to yield to the light.

«- Do you say that God has both sexes, oh Trismegistus? - Not only God, Asclepius, but all animate and inanimate beings. It is impossible for any entity to be infertile. Because if the fecundity of all existing beings were removed, it would be impossible for them to always be the same as they are. I for my part say that, by nature, Mind, Nature and the World contain in themselves the power to generate and preserve all things that have been born. In truth, both sexes are filled with the power to procreate, and the mutual connection of both, or better, the union of both is something incomprehensible, and you can already correctly name her Cupid or Venus or both names at the same time. I want you to keep well in your mind what follows, the most true and evident of all truths: the Lord of all Nature, God,

invented and granted to all beings this mystery of eternal procreation, whose natural attributes are the utmost affection, happiness, joy, desire and divine love. And it would be necessary to explain more how much is the force and the imperious necessity of this mystery, if it were not well known to each one, in their intimate feeling, from their own experience. Because at the extreme moment of orgasm, which we reach after repeated rubbing, when one sex pours its seed into the other, you will notice that each one greedily snatches up and hides in itself that of the other, and that at that moment, through interpenetration mutual, the female seizes the strength of the male and the male abandons himself to the languor of the female. Where the act of this mystery, so sweet and necessary, is carried out in private, lest the mockery of the ignorant vulgar embarrass the divinity of both natures during sexual union, and much worse if one is exposed to the gazes of impious." (From Hermes Trismegistus addressed to Asclepius, verse 21. Corpus Hermeticum)

V.
IN THE BEGINNING

GENESIS 1

"If your heart is a volcano, how do you expect flowers to sprout?"
Khalil Gibran (1883 - 1931)

And after the foundation [of the world] Saclas said to his angels:
« *Me], I am a Jealous Di[o] and apart from me no other exists»,*
since he believed in its reality. Then a voice came from on high saying:
"There is Man and the Son of Man", because of the descent of the
image from on high, which is similar to his voice on the height of the
image that he has seen. Through the vision of the image from above,
the first creature was shaped. Because of this there was repentance. He
received his fulfillment and his power by the will of the Father and his
agreement with what he accepted from the great race incorruptible,
unbreakable, from the great powerful man of the great Set, so that he
would sow it in the aeons that have been begotten so that by him (=
repentance), the deficiency will be completed. Because he had come
down from above to the world that is the appearance of the night.
When he arrived, he prayed jointly for the seed of the Archon of this
aeon and [the] authorities that existed from him, the polluted one that
will be destroyed by the demon-engendering god and prayed for the
seed of Adam, which is similar to the sun, of the great Set. **»** (1st Set
or the Gospel of the Great Invisible Spirit 1:58-60) The origin of
man is a lurid subject on which we will only make a brief reference,
since we want to properly address it in the following books. We
hope this is the case, with God's blessing, but first I consider it
necessary to summarize the aspect of Creation, especially according
to the biblical and parabiblical versions. We are understanding that
the appearance of this planet and of man were not by chance and it

was expected that they would eliminate disability, restoring what was theirs to the Holy Spirit. Hasidism explains that the definition that translates into Spanish as "does" also means in its Hebrew root "rectifies", as it is said at the end of the story of Creation (the seal of the Seventh Day, Shabbat): «the one that *Elohim created to do* », referring to "do" in the sense of "rectify". The Jewish sages hold that this means that God gave us the task of completing the rectification of His Creation, but if it is perfect, why does it need "rectification"? What Sophia began with Christ would continue with the lineage of Set and would be sealed with the coming of the Christ in the flesh. Precisely this text -quoted above- from Set mentions the fact that Sophia «*had descended into the world that is the appearance of night*», increasing the chances that she would illuminate the chaos with her presence and differentiate between the ignorance of the archons and their creation of what is the truth of the Eternal Realms and its light. That light came to this world, as Juan said, but it began with Sofia and continued with the lineage of Set, since this same lineage " *is similar to the sun, of the great Set*."

Muhammad apparently commented: "*Your Lord is God, Who created the heavens and the earth in 6 days. Then, he has installed himself on the Throne. It covers the day with night, which quickly follows. And the sun, the moon and the stars, subject to His order. Are not creation and order His? Blessed be God, Lord of the universe!*» (Quran 7:54) All these mythologies, and even certain religions of yesterday, remind us of principles mentioned in the Genesis of Moses. Thus, we have to inquire as to the meaning of the first chapters of this book of the Torah. Many fans and experts in submerology support the theories of the Russian researcher and writer Zecharia Sitchin, especially on this subject, exposed in detail in his work "Revised Genesis". Sitchin argued that the Genesis story and the Sumerian Enuma Elish were closely related to the terra-formation of our solar system, both being the same tale but modified by the passage of time. Although, the writing of this first manuscript is attributed to Moses, just like the rest of the books of the Torah, the Jubilees, the Revelation of Moses and the book of Job. What does the first text of the Tanaq (Old Testament) tell? Does the first chapter of Genesis relate the beginning: the creation of the universe, of planet Earth, of man, etc.? It is possible that we should reorganize the ideas created on this text, leaving behind the postulates that we have been taught. The book of Barashit -in Hebrew- or Genesis (Greek word that would translate: "generations") seems to basically express the development of the thought of the gods and the intervention of the Holy Spirit, whose result would be what exists. This is known today as creationism. Even with everything, said work seems to have certain deficiencies and contradictory nuances. This is where we should undermine.

Moving back a bit, we can elucidate mythological creationism, which has a certain resemblance to the genetic, Barashit affirms that the "gods" (Hebrew voice: "Elohim") in the beginning created the Heavens (Shamaim) and the Earth (Aretz), which is exactly the same principle that the old legends expose. What few stop to think is that it is not explained how it was created. The reason may be that the Hebrews kept the tradition of the prophet and scribe Enoch, where the subject was already exposed, in addition to the fact that Moses also detailed it in the Book of Jubilees. It is exposed that in said principle the Earth was already "disordered and empty", a quite inaccurate translation, taking into account that in a vacuum there is nothing, and "things" are those that, in an inappropriate place, generate disorder, therefore that since there was nothing, what disorder was there going to be? «*Who visited the earth for him? And who put the whole world in order?*» (Job 34:13) The Hebrew "Tohú" and "Bohú" are striking words, since "Tohú" is "vanity" or "empty" (examples in: 1 Samuel 12:21 and Isaiah 29:21) in a concept of translation, while in another it is "chaos" (ie in Isaiah 24:10) and in another it is something "in vain" (see Isaiah 45:19). It cannot be empty if it is chaos, since chaos is SOMETHING critical and emptiness is simply a "nothing". Tohú is singular from the plural word "Tehom" (Abyss), which takes us back a few steps? Tehom is also understood as the Primal Ocean from which the ordering of the Earth was produced. It is also a term to describe the depths of the sea, the deep waters (example: Exodus 15:5) or it is simply understood by the Abyss or source of subterranean waters, as exposed in Deuteronomy 8:7. Let's see that the word "Tahalá", which comes from the same root, also speaks of "need", "lack" or "emptiness", being translated like this in Reina Valera, but in other texts compared to Aramaic, it is defined as " mistake". This is how it is seen when Job implies that God "... *does not trust his servants, and noticed foolishness/error in his angels.*" (Job 4:18) With this he confirmed the words of Enoch, when he affirmed: "*Now the angels of Heaven are guilty of sin and your wrath falls on the flesh of man until the great day of judgment.*" (1 Enoch 84:4)

In etymology, the Hebrew expression "Tóhu va-Bóhu", which translates as: "without order and void", is a case of endiadis, in which "Bóhu" does not specifically mean "without order"; rather, both words, "Tóhu" and "Bóhu", jointly express the idea of Chaos. Starting from "Bohú" we find the voice "Bhl", "Bahal" or "Bohal", which is to be terrified, to be disturbed, to be intimidated, to remain paralyzed with fear, or simply to "hurry up". Also from there derives "Behemoth" (the Beast), this being a feminine plural. So others would have translated that the Earth was "empty and terrified", but still, if there was nothing, what was terrifying and who was it terrifying? The most daring recent Sumerologists say that the primordial Earth, before being Eridu (Aretz, in Hebrew) was an enormous watery planet that split in two and whose name was Tiamat, a word whose root is similar to Tohú, and that in the myth Sumerian-Akkadian represented the "dragon goddess" who would be mother earth. Adding all this together, we could understand the postulate of Edward Stone, from the California Institute of Technology (Caltech), the chief scientist of the Voyager program, who argued that "cosmic accidents were powerful sculptors of the solar system", since the Mesopotamian tradition would *be* an allegory of a space cataclysm, a cosmic event that would have resulted, not only our planet, but the entire solar system. But if the events were so remote, how did the Sumerians or the Hebrews know about them? But how did the prophet Moses know about these events that occurred, according to classical creationism, at least 8,500 years before him? If the book of Barashit was written about 3,500 years ago, then it is not very new, since we find its main facts included in the "Enuma Elish" (Sumerian Creation epic), which dates back about 5,800 years, and others are described in the Atrahasis epic. Someone inspired Moses about this narrative and told him about its intricacies, although the story itself is mentioned by many other peoples with different nuances. In addition, the Hebrew script appeared "encoded", and Moses was neither a mathematician nor did he have computers capable of carrying out such a prodigy of computerized latticework, which is this compendium of Hebrew literature that we call Torah (in 1995 it was discovered that the TANAK was mathematically encoded and even in those days it was practically impossible to reveal all the secrets encrypted in its lines).

So having drawn some highlights in this regard, let us summarize that the Scripture teaches us, in the Hebrew language, that « *in the head the gods made, that one, the Shamaim and that one, the Aretz.* » (Barashit 1:1) On this let us note:

A. Barashit (Bar-esh-it) means in Aramaic: "Son-fire-placed", which in symbology means: "*That [was the] result of [divine] fire*", or in terms of the fact that life took place because "*That was the premeditated fruit.*" It is also broken down as "Ba-rosh-it", which means: «*in the heading*» or «*that as a heading*». In the same way, the single word "barashit" translates: "*that creation*", or also "bará-shit", which means: "*done what was imposed.*" In short, no matter how it is translated, it docs not literally imply a relationship with the beginning of everything, but rather with a work developed in what is already established, or working with ready means in hand: the beginning of the development of the material world having gone through a previous chaotic event.

A. Elohim (EL-oh-im) means: "*God-those-[who]-mourn*" or "*God-those-[who]-support*". That is to say: «*God of those who lament*» and also «*Gods who support*». Even so, and although the Jews do not want to see this observation, this word translates and is applied generically as "gods", or simply and simply refers to the "divinity" or the "deity" in a global way. If he were referring singularly to the Most High God, he would say, on the contrary: "El" (God), "Eli" (my God), "Eloah" (the God) or "Elyon" (great God), as in other Common Bible references. In any case, the word "Elím" (Gods) is not used either, but instead we have the initials "H" (hei) in the middle of the word, which broadens the meaning of the word by adding "exaltation". The "He" or "Hei" is the fifth letter of the alphabet, which symbolizes the divine and is associated with the sacred. Hei is equivalent to celestial support and sometimes expresses supernatural intervention due to human lamentation. Said spelling also implies that oneself must "take" what belongs to one to help others, according to

Hasidism. Likewise, Hei means: breaking, taking seed, contemplating and revelation. Now, the Scripture says that there are other "gods" (Elohim), written in the same way. This problem argues that there are many "gods", and for this reason it has been imperative to reveal the intricacies of who God was in that narrative, what other gods were related there and if there was a mixture of entities at that beginning, which is not clearly expressed. Genesis – although it is hinted at in the Jewish Haggadah. Let us keep in mind, in any case, that in case a "deity" is identified with "Yehovah", it already clarifies which side it is on, that is, it is understood that it is the "deity" known as Jehovah: "Yehovah Elohim", or translated as "Jehovah God" or "Jehovah the Lord".

A. Aretz: Derived from the Sumerian word "Eridu", which was the first location or civilization on Earth when the Anunnaki arrived (it is a linguistic transformation of the Sumerian E.RI.DU, which means "home to go from afar", called mainly HA.A.KI (place of the fish-water) in Sumer). It is understood that they established it themselves. It is also associated with "Eretz" (land, terrain, territory or nation).

A. Shamaim: Derived from the Akkadian word -and this, from Sumerian root- "Shemaim", which means: "People of the Rocket" or "Nation of the Launch [upwards]". In the Aramaic language the word, in its etymological root, is broken down as "Sham-maim" or "Shama-im", which means: "There Peoples", "Above Nations", "Towards There [there are] People". The very final addition, "im" (masculine plural), is also applied as "iam" (sea), which comes to be a denominative form for the proliferation of crowds and people, used symbolically in the vast majority of biblical concepts.

According to the theory of the Jewish scholar Félix Guttmann Van Katz, in this context we can assume that nobody was talking about the creation of the world, as taught in classical Creationism, but about two important establishments: one on Earth, which would be the first civilization established by the Elohim, and which would give rise to the cradle of humanity and the spiritual and military base of God's operations among men; the other would be the air base where thousands of people would work on heavenly affairs and the problem caused by the rise of evil. Not by chance, the Enuma Elish says this: that places to rule on Earth (Aretz-Eridu) and an establishment or space base in heaven (Shamaim) were built. In fact, the Babylonians tried to emulate Elohim's shuttle in Mesopotamia: the Tower of Babili or Babel. We could reflect on this theory, although Zecharia Sitchin also points out some points to take into account, and which I will highlight later.

Going back to the point of chaos and desolation, was the "Earth" establishment in chaos and desolation as a result of what had been happening? Was "darkness" (ignorance) in, or dominating, "Tehom" (Abyss or cliff)? So there is an affinity of some kind between that abyss of Hades that was in ignorance and spiritual darkness, with the situation of universal truth, a time in which the forces of evil exercised power or authority in the Underworld. So that the problem of humanity must be due to those "forces" that, being in "darkness", influenced or acted from the "abyss", an aspect similar to that already proposed for at least 2 decades by the teacher Félix Guttmann Van Katz. The point is that the proto-Hebrew prophet Enoch (whom the Sumerians associate with Emedurana and the New Age metaphysicians associate with

Metatron), wrote: «And the *Lord summoned me, and said to me: Chanoj, sit on my left with Gabriel. And I knelt before the Lord, and the Lord spoke to me: Chanoch, dear, all that you see, [about] all things that are standing firm finished, I speak to you, even [about things] before the primordial beginning, everything that I created from no being, and visible things from the invisible. Listen, Janoj, and take these my words, because not even my angels have I told my secret, and I have not told them [the origin of] their departure, not even my endless kingdom, they have not even understood my creation, which I tell you today For before [that] all things were visible, I only used to go about the invisible things, like the sun from east to west, and from west to east. But even the sun has peace in itself, while I find no peace, because I was creating all things, and I conceived the thought of laying foundations, and of creating visible creation."* (2 Enoch 24:1-5)

One of the most suitable people to share his knowledge about the Creation of the world is this great scribe of the Lord, the prophet Enoch, nothing more and nothing less than Noah's great-grandfather. He, in this sense, left many valuable texts from which we can feed and acquire, not only knowledge, but wisdom. Enoch wrote that the Lord said to him: "*And he came out of the highest [things] because a visible came out of the invisible and Uriel came out very powerful. And behold, his appearance was a great light without comparison. And I told Uriel to give birth and let your birth be visible. And she gave birth and came out great, and I was in the middle of the light. And the light fluttered, and out of the midst of the light [a] great world, a vision of all Creation that I said was created. And I saw that it was good. And my room [was a] throne and I sat on it. And to the light I said: go upstairs and sit next to the throne and there was a foundation for lofty things. And there is no beyond the light. And then I turned and looked from my throne and I called a second time ...*" (2nd Enoch 25:1-6) After this first creation that the Almighty developed through Adoil -which is clearly oriented to the appearance of his great dwelling place and heavenly kingdom-, he continued doing more: «*And I summoned the very low one a second time, and I said: Let Arjas (spirit of creation) come out hardness, and he manifested hardness-solidity from the invisible. And Arjas came up, hard, heavy, and very red. And I said: Open, Arjas, and let [that] be born of you [all of it] there, and he undid, [so] an era came forth, very great and very dark, bearing the creation of all things low, and I saw that it was good and said to him: Go therefore down [far below] below, and [manifest] until [make] yourself firm, and be a foundation for low things; and it happened and he went down and repaired himself, and the foundation came for the low things, and under the darkness there is nothing else .*" (2 Enoch 26:1-4)

After creating light and physical things, it seems, the Almighty creates water, in what is barely the first interval also cited in the book of Genesis: «*And I ordered that there should be taken from light and darkness , and I said: Be dense, and it was so, and I parted it out with light, and it became water, and I parted it out with darkness, below the light, and then I made firm the waters, it is say the abyss (unfathomable, bottomless), and I made a foundation of light around the waters, and I created seven circles from within, and imagined the waters as wet and dry crystal, which is to say as glass, and the circumscription of the waters and of the other elements, and I showed [to] each of them their way, and the seven stars each of them in their sky, [because] they go like this, and I saw that it was good. And I separated between the light and the darkness, which is to say in the middle of the water here and there, and I told the light that it must be day, and the darkness that it must be night, and it was late. and tomorrow was the first day.*» (2nd Enoch 27:1-2) It is quite obvious that the Most High made use of "light" (potential, energy, spirit, force) to cause changes in the development of things, since it is a symbol of evident action of his fellow servants, such as Christ and Sophia. On the other hand, what would those "seven circles from within" be? And why does he speak of "seven stars"? Do you associate the seven circles with those seven stars and "their heavens"? This seems to inquire that the circles are the dimensions or Rakiot (Heavens) mentioned in the Talmudic traditions or that they are the 7 main angels administering each of these 7 places, an aspect that launches us to the obvious explanation that it speaks to us of our companions. spacecraft and their orbits. Consequently, the seven stars may be the Seven Spirits of God or archangels that administer these seven "dimensions" and their waters, but specifically referred to here would be the planets that share this system with us. When you say waters, do you mean worlds, civilizations, races, humanities or sub-worlds? From the book of Revelation (chap. 17:15) we know that the mythical meaning of water represents multitudes of people, although water is also a primal substance of life.

Now, the Creator speaks to Enoch of another fundamental part of this manifestation of his work in matter, telling him: «*And then I made the celestial circle firm, and I caused the low waters which are under the sky to collect themselves, together, into a hole, and that the chaos became dry, and it became so. Out of the waves (wave) I created hard and great rock, and from the rock I filled the dry, and the dry I called earth, and the middle of the earth I called the abyss, which is to say the bottomless (unfathomable), I collected the sea in a place and limited together with a yoke. And I said to the sea: 'Behold, I give you your eternal limits, and you must not break [,] let go of your component parts. So I quickly made the firmament. [A] this day I called the first-creation."* (2nd Enoch 28:1-4) What did he mean by making the circle of heaven firm? Here he does seem to be describing the creation or "construction" of planet Earth and/or something else outside of it, including the celestial mold: "*Know that this Earth is nothing more than a portal, guarded by powers unknown to man. However, the Dark Lords hide the entrance that leads to the Skyborn land. Know ye, that the way to the sphere of Arulu is guarded by barriers opened only by the man born of Light."* (Table 14. Emerald Tablets of Thoth) This appreciation even supports the theory of the Hollow Earth and the underworld. Likewise, we can see that it follows the line of thought that suggests that the concept of "Sea" refers to a specific place of creation, outside the Earth, perhaps even in, or outside, our solar system, at the same time. that identifies the world's oceans (many cases of "revelation" about "evil" and its forces can be located outside or far below the ocean floor). At this moment we see that he does not say that this was the Second Day or that he spoke again of the First Day, but rather describes said event as "the First Creation", still within the first interval of time. The Israelite king David wrote: «*Praise him, heaven of heavens, and the waters that are above the heavens.*» (Psalm 148:4) Heaven of heavens? Waters above the skies?

In the Mayan culture, deities such as Tepeu, Gucumatz or Kukulkan and Hurakan were related to the beginnings of the world. The most important gods of this civilization are said to have raised the earth that was submerged under the waters. Specifically, it is supposed that Hurakan or Huracán lived in the windy night above the waters and repeatedly invoked the "earth" until it definitively emerged from the seas. We see another example from Mesoamerica in the Teotihuacán culture, where Tláloc originally represented terrestrial water, while, for its part, the feathered serpent (who centuries later became known as Quetzatcóatl), celestial water (here we see once plus the concept of upstream and downstream). This idea continued even after the fall of the city, passing to Tula, and from there his cult spread among the Nahua peoples. In Sumerian mythology, let's remember that mother Earth, Ninhursag, is also spoken of, which is the best-known name for Ki. She, it is said, was the Earth and mother goddess who usually appeared as Enlil's sister, but in some traditions was his consort. Officially she was born from the union of An and Nammu, although sometimes she appears as the daughter of Kishar (all of these correspond to the Sumerian names of the planets of our solar system). In the early days Ki/Ninhursag is supposed to have been separated from heaven (An), and taken out by Enlil. The peoples of the past would have erred believing that their patron gods were the cause of these geological events. It is not strange, then, that characters of those times such as Moses and Enoch gave directly heard versions of the true nature of these events, thus leaving behind so many inconsistent or incomplete legends that detracted from those who really deserved it and motivated more misguided polytheism.

The Creation

The Apostle John wrote: "*In the beginning was the Word, and the Word was with God, and the Word was God. This was in the beginning with God. All things were made by him, and without him nothing was made that was made. In him was life, and life was the light of men.*" (John 1:1-4) What Word is the missionary John speaking of? Whatever that Word was the reason, motivation or force that created everything that exists. John says that "in argi" (in Greek: "from eternity") the Word already "was" or "was", that is, the Word of God -not the Bible, but the power of his "order"- was the first thing that existed. Starting from the famous word "logos" existence emerged. According to it -the Word-, which existed with God, and in itself was God -whether as a representation, as Himself or as part of his Assembly-, all things were made, and if it were not for the Word, nothing that has been made would have come into existence. In the Word was Eternal Life, and this was the "light" that guided men in the universe, the one that "illuminated" them, and the one that from Christ brings lucidity and truth to those who learn from him. God said to Enoch: "*If I turn away my face, then all things will be destroyed.* » (2nd Enoch 33:5) More of this information is in the apocryphal book of John (Secret Book of John) and in the apocryphal of Seth (Gospel of the Egyptians or of the Great Invisible Spirit). And what is that of the "light" that guided or "illuminated" the men of the cosmos? That light that manifested the Word of God came in the person of Jesus: « *That true light, which enlightens every man, came into this world. He was in the world, and the world was made by him; but the world did not know him. He came to his own, and his own did not receive him. But to all who received him, to those who believed in his name, he gave the power to become children of God; Which were born not of blood, nor of the will of the flesh, nor of the will of man, but of God. And that Word was made flesh, and dwelt among us (and we saw his glory, glory as of the only begotten of the Father), full of grace and truth.*" (John 1:9-17) This implies that the universe was created by the Word of God and that same Word was the light that was manifested to us in the person of Jesus, carrying out his ministry or work even though his people, Jews, at the time they would not have assimilated it.

« *You alone are Jehovah; you made the heavens, and the heaven of heavens, with all their host, the earth and all that is on it, the seas and all that is on them; and you give life to all these things, and the armies of heaven worship you.*" (Nehemiah 9:6) IHVH is credited with creating heaven and, furthermore, the heaven of heavens. It is also assumed that with this establishment came the corresponding hosts with the direction of all the things that appeared. Likewise, it is said that He gave rise to the planet Earth, to the seas and to all forms of life, including, likewise, the fact of maintaining the life force of all creatures and beings that swarm around. Now, we cannot assume that this light -of which John spoke- that illuminates every man was already here, because if not, why would they bring it? And if it means that this light now acts in humanity thanks to the fact that it came before, why is the world in such a bad way? Obviously, that illumination of which he speaks was brought from another place where its inhabitants do live in it. Through Christ Paul affirmed: *"He is the image of the invisible God, the firstborn of all creation. For in him all things were created, things in heaven and things on earth, visible and invisible; be thrones, be dominions, be principalities, be powers; everything was created through him and for him. And he is before all things, and all things exist in him ..."* (Colossians 1:15-17) So, it is true that Jesus was predestined to be King of the Universe, and everything that exists was prepared for this task. Even John is emphatic when he wrote that "the world was made for Christ's sake." Therefore, we understand that the world (cosmos) was "established" because all men hoped to see the only begotten (only of their genetics) and firstborn (first of their genetics), because this was the corporeal image of the invisible God, even though he was not God himself – like the Father. And so it was done by God, as King David said: «*For he spoke, and it was done; He commanded, and it existed.*» (Psalm 33:9). Although, Paul also referred, about the Heavenly Father, who was manifested in visible things although He is not visible: «*... For the invisible things of Him, his eternal power and deity, have been made clearly visible from the creation of the world, being understood through the things made ...*" (Romans 1:19-20) That is, the Father manifests himself in his creation: we can see him through all the things that exist in the universe, and in the case of the "qualities", are observed through Jesus Christ – while in person it is done by the Ancient of Days.

But what is the world? The word "world" comes from the Greek: "cosmon", which means: "living universe", and from the Hebrew: "olam", which means practically the same thing: "universe of the living". So, what we call "world" as a planetary sphere, writing and ancient language refer to the cosmos full of all kinds of life forms. The same code that called itself "Son of Man", because a "son" is the "result" and "fruit" of a thing, and in this case it is "the result or fruit of the Adam code", a creature special. The gender Adam (A-DM = God-Blood) is the genetic code of God, not because the Father has blood or a genetic code, but because it implies that we are gods and men at the same time, and that we have a link with divinity. Jesus taught what life was like in the cosmos before the Sakla Rebellion, Satan and the other fallen angels. Because of this sedition, Jehovah carried out a purpose of Salvation, first with a nation and then with the whole world that accepted the Son of God –although there was predestination in said matter. Jesus teaches this in a parable, in which he affirms that the Kingdom of Heaven is similar to the Son of Man, who sowed the genetic code of the Sons of the Kingdom in the Universe; but while they were naive, the Devil came and sowed tares (Sons of Iniquity) between the good and everything made in creation, and then he left. When his actions were successful, it was seen that in the fruit of the children of the universe there were good and bad, and there were good and bad consequences. Then the servants of the Almighty Father came and said to him: *«Lord, did you not sow good seed in your field-universe? Where, then, do you have tares?" And He said to them, "An enemy has done this." And the servants said to him, "Do you want us to go and pluck it up?" He said to them, "No, lest when you uproot the tares, you also uproot the wheat with it."* The Father wanted to avoid bloodshed and deliberate judgment. He did not want to act without first knowing the disposition of each one's heart and where their works were leaning. So the Father told his servants to let everything unfold by itself until the *"End of the Age"*, which is the time when he will send his angels and cast those who cause offense and iniquity into the lake of fire. (Matthew 24:43). The same thing that John said about Christ was reiterated by another contemporary: *"By faith (visualization) we understand that the Universe was constituted by the Word of God, so that what is seen was made of what was not seen."* (Hebrews 11:3) Then there was "something" that was not seen, and from where everything came when it can now be seen. So what today is manifest of Creation comes from a previous world or universe of which this is a spitting image or replica, a type of initial design that resulted in our universe, although the previous one is perfect and superior.

Unfortunately, practically all the sources that speak of the universe before the Sakla Rebellion are not compiled in the Bible. Almost the majority of them constitute a large part of the ancient writings kept in museums, large libraries, medieval churches, in the Vatican, they were found recently or simply "put aside" (canonized). But having access to some texts recovered in Nag Hammadi (Egypt) and in Qumran (Israel) in the mid-40s, we can synthesize that "eternal life" was the "light of men" who lived in the cosmos: terrestrial, intraterrestrials, extraterrestrials and ultraterrestrials. The prophet Baruch wrote, for example: "... *Or will the world return to its nature before-time, and return to the silence of [the] primeval age?...*» (2nd Baruch 3:6) He was knowledgeable in such matters and spoke about the nature of the world "before time", that is, before the times were established; the same was primal that was in the "silence". He also said: « *[God], you show great acts to those who do not know; [you] who break the box-structure of those who are ignorant, And the lightest is dark, And [you] reveal what is hidden from pure matter..."* (2nd Baruch 54:5) Therefore, there are "realities" transcendental things outside the level of "matter", "hidden" things that only God makes manifest. Everything that exists came from silence, from the invisible, from the Mind of the Creator, from the world of Thought, and manifested itself in the visible, in the dimension of matter: «Then a Thought manifested itself in a silence *and a living silence of the Spirit, and thus a Word from the Father and [this was the] light."* (The Great Invisible Spirit, Gospel of the Egyptians 5:58-59. Nag Hammadi II Library) With this proceeding the "image" of men appeared: "Then came the great living self-generated [Word*], [the] true God, the unbegotten nature, [...] who is [the] Son of the [Most High,] Christ, who is the Son [of] unspeakable Silence [who] came into being from the great [invisible universe] and incorruptible [being]. The [Son of Silence and in] Silence manifested himself."* (The Great Invisible Spirit, Gospel of the Egyptians 5:60)

Ezra the scribe recounted: "*He said to me: "At the beginning of the circumference of the earth, before [setting] the portals of the world in their place, and before [the] assembly of the spirit flew away, and before the noise of thunder sounded, and before [the] flashes of lightning flashed, and before [the] foundations of paradise were laid, and before [the] beautiful flowers had been seen, and before [the] the competitions of the movements had been established, and before [the] innumerable legions of angels were assembled, and [before] the heights of heaven were raised, and before the measures of the firmaments were named, and the feet were created before Zion ..."* (4th Ezra 6:1-4) Let us note how great a knowledge of the Scripture Ezra had, considering that he knew well how the High and Exalted founded the foundations of the world and even of the heavens. that God speaks in different ways, one of which is symbols. Jesus presented himself as the light (John 8:12, 9:5 and 12:46), and was thus defined by John (John 1:9 and 3:19). Although, we understand that light represents clarity, clarification, truth, response, guidance, and other similarities. However, the biblical story of light can already be found in the book of Genesis, chapter 1, sometimes seen as a metaphorical or allegorical idea of how God intervenes in his Creation to clarify the things that the evil one has obscured. However, the light that came to the world as a revelation of God's truth was manifested with Jesus, so the literal aspect of the Genesis narrative must be taken seriously, especially under the magnifying glass of investigation that interests us. Even with everything, it is evident that sunlight does not seem to be what is defined there, since this flame or initial brightness enters on the First Day of said narration while, supposedly, the sun takes place only on the Fourth Day.

The writing of the first book of Moses, wields: «*In the heading Elohim created that Shamaim and that Aretz. And Aretz was in chaos and desolation, and darkness [was] over the face of the Abyss, and the Ruach Elohim fluttered over the face of the waters."* (Genesis 1:1-2) Shamaim we translate as Heaven; Aretz we translate it as Earth; Ruaj we translate it as Spirit; and Elohim we translate as God (despite obviously being a plural). Initially it opens the narrative highlighting that in a "beginning" what we define as Heaven and Earth was created. What sky is that? The universe? Well, the Earth is already part of the universe itself. The atmosphere? Well, the atmosphere belongs to the Earth. The ethereal sky? Well, what about the rest of the cosmos? Is there no mention of its creation? It is necessary to perceive which sky is referring to, which came to be "created" precisely at the same time as the Earth itself, without including other planets, nebulae, galaxies, solar systems, asteroids, comets and other things that make up and populate space. It would also be strange for Elohim to create the Earth based on a chaotic mold if God is pure, right? Ergo, where would the initial mass come from to form the Earth? Did he create it out of nothing? If he created it out of nothing, why would it be in chaos? What is comprised of "chaos"? Then appears such a "darkness", which is above an Abyss. Why only over that abyss? And where was that abyss located? What was the point of mentioning that aspect of Darkness and the Abyss? On the other hand, they speak of Elohim and also of Ruaj Elohim, in this case moving from one side to the other over some water, but what difference is there between Elohim, Ruaj Elohim and Jehovah Elohim? And to all this, where, without further ado, was the water going to come out? It should be openly noted that all of these points are largely overlooked by creationists and even mere Bible readers.

In the Book of Jubilee, also attributed to Moses, the genesis description of the first aeon appears summarized as follows –according to the acquired translation: «*For the first day that created the heavens and that are above the earth and the waters and all the spirits that serve before him: the angels of the presence, and the angels of sanctification, and the angels [of the spirit of fire and the angels] the spirit of the winds, the angels and the spirit of clouds, and of darkness and of snow and hail and hoar frost, and angels of voices and thunder and lightning, and angels of hot and cold spirits, and winter and spring and autumn and of summer and of all the spirits of his creatures that are in heaven and on earth, (has created) and the abysses of darkness, [eventually] (and night), and light, dawn and day, which He has prepared in the knowledge of his heart. And then we saw his works, and we praised him, and he praised him in view of all his works, of 7 great works that He created on the first day.*" (Jubilee 2:2-3) It is necessary to consider that the process of formation of living things, whether inert or moving, began with Heaven and Earth, followed by maritime deposits and then "the spirits that serve them." Why are these cooperating "spirits" not mentioned in other references? Or are they there and we haven't glimpsed them? In fact, they are specifically indicated under their functions and in their own genre. Here it is relevant to analyze that the angels (in Hebrew "malachim", in the plural) are really "messengers", since the word "angel" is not Hebrew but comes from Greek mythology and with a weight and value different from the Biblical malaj. For this reason, the malaj is considered primarily "spirit", not as an immaterial entity but as an existential form of a spiritual nature, oriented towards matters relating to heaven. If we focus properly, we will see that the "spirits" identified there are, in fact, what the Bible calls an angel, but other traditions call them differently, often considering them "spirits" and "geniuses" (I am not saying with this that angels are geniuses, but there are many types of non-human entities that are entirely "excellent" and of a different essence from ours, which govern the elements and laws of the world).

Consequently, the same thing that we read in previous pages about the appearance of angels, spirits and others, is what we see here, but being God who attributes the formation of these, since it never gives rise to chimerical questions about the "apparent", but assuming that He is really behind it all. Hence the matter of talking about the angels who govern all that, either as spiritual beings or directing other spiritual forces linked to the observant powers of Creation. That applies to the elements, whether in Heaven or on Earth, and further on it also includes the description of the dark abysses -which clearly were consolidated with the location of things-, continuing the development of all this with the separation of light and darkness, which he defines as incorporated into an age, the first. In this way, although the elapsed or founded aeons were, in principle, seven, the number one itself also included seven great works in it, once again exposing the work of counteracting negative patterns and establishing holy foundations on them.

The Light

The Levite deliverer of the Israelites, Moses, noted: "*And Elohim said, Let there be light, and there was light. And Elohim saw that that light was good, so Elohim differentiated between light and darkness. And Elohim called the light, Day, and the darkness, Night, and it was evening, and it was morning, the aeon one.*" (Genesis 1:3-5) The next aspect that Elohim performs is determined by a verbal command. That ongoing determination is defined in the

Hebrew language as "or" (light), which symbolically represents "information" and "truth". At this point the deity seems to order that illumination exist, but the Fourth Day is when the greater light, the lesser light and the stars are seen. So what is that first light? Another thing is that it creates that light and separates it from darkness, an action that is quite absurd, since light itself is divided from darkness, the two cannot occupy the same space. Paul wrote: "*For God, who commanded light to shine out of darkness ...*" (2 Corinthians 4:6) That flame that enters the dark area illuminates it, but it is not a star. According to common ideas, this would define representative aspects, alluding to the Kingdom of "light" and the Kingdom of "darkness". So calling the light, then, as "Yom" (Day), identifies that process or step as a period of time, since Yom comes from the word "aeon" (era, age, century). If an aeon is defined as an age or era, and also as a creation or existence, the 6 days are best stated as 6 ages. For its part, figuratively, ignorance or evil was described as "Laila" (Night). We can see that Elohim begins to classify developed and fulfilled objectives under the plural name of "Yomím", beginning with the distinction between truth and lie, between clarification and ignorance, between the children of light and the children of the darkness. The Hebrew word "or" (light) is made up of 3 letters: A (Alef), V (Vav) and R (Reish). Alef represents the initial, the divine, basically identifying the connection between God and man. Then Vav represents man, also identifying the division or connection between two points, usually to separate the light from the darkness or to restore to the light what has been taken away. On the other hand, Reish refers to a beginning, leadership, or head.

It is necessary to keep in mind these 3 principles that make up the identifying weight of Light: divine intervention to change a dark consequence into luminous action, restoring what was taken by evil thanks to the introduction of man, who with the help of

God, would intervene as semi-divine beginning a mission under his responsibility and leadership. It is now worth bringing a phrase from Thoth, which says: « *The Light is life, since without the great Light nothing can even exist. Know that in all formed matter, the heart of Light always exists. Yes, though bound in darkness, the inherent Light always exists.*" (Table 10. Emerald Tablets of Thoth) It is obvious that Moses, in his cosmogonic account, when referring to a "day", did not speak of a period of 24 hours, because the day begins, or, from 12 o'clock in the night from point A until midnight of the following "day" (point B), or 12 hours, if it is from the time the first star rises until it sets (starting from Ecuador). But here it is assumed that it came to be from "afternoon" to "morning", that is, "during the night". It cannot be a day during the night, so we understand that he speaks in symbology. Did Elohim begin his work while the satanic system (night) exercised its power? Or was he just working in the dark? Numerology, Kabbalah, symbology and etymology have played an important role when scrutinizing the Scriptures with greater possibilities of finding new or more advanced data - although there are those who use it for occult purposes - as expressed by the Zohar. Using a numerical order of 10, as suggested by the Hebrew-Aramaic biblical code, we find, from verses 1 to 6 -without going any further- in this first chapter of the book of Genesis, an exposition that "in that which was, [*as it was] nothing of what was made would have been made but by their mouths* ». This is decidedly very similar to what was expounded in John's letter in chapter 1. Later it can be drawn: "*[they] who left them to their desires,*" or "*who cheated their desires.*" Was there something planned in advance? Did you know what was going to happen? Who was left to their desires, so that they believed that they carried out their own will? The gods of chaos? Let's keep in mind that we talk about things before the world as we know it existed. Well, with hindsight you can get something that means: "

and their tongues were all made to turn into their blood." That is, did their own common ideals, plans, and goals lead to their own death and destruction?

Now, using the Sumerian number system, which instead of being based on 10, is based on 12, we also find interesting things: "*There was cruelty of a foreign community, a town that generated bitterness*", or we can also translate it as: "*Those who were angry were [those who turned their] backs, and they brought bitterness to the people.*" They abandoned the kingdom of heaven, to worship the Beast (Romans 1:25). It is then decrypted: «*so how did it prosper? God brought wine.*» That is, how did you get out of trouble? How was the matter resolved? The Most High brought anointing (the symbolism of wine has various connotations: anointing, blood, rejoicing, debauchery...). With another analysis of the Barashit (name of the book of Genesis in Hebrew), without leaving the first chapter, and in this, its first verse, we find more hidden things. With the cabalistic systems of Notaricón, Temurá and Gematría (the three systems of study of Kabbalah) we can get this composition: BT (house) BA (to come) AM (yes... mother) AT (that) HM (people, masses) VT (and brand, stamp) HTZ (the just one). Although this is more extensive to expose than how I define it here (only putting a word as an equivalent example of said two Hebrew letters), it can be understood that the idea was to establish a home or dwelling so that the beginning of the life of the people would come. or people already alive, which symbolize a kind of "holy lineage". With another numerical system, from top to bottom, we get: BBAAHVH, which takes the Bet, the Aleph and the Hei repeated twice, together with the Vav, once, and between two Hei. The reiteration of the Bet reinforces the idea that the planet was not created by chance but that it was already decided that it would be placed as a habitation for something sublime (Aleph), which is also reiterated twice. To this is added the expression, thought, speech and action (Hei) that cover the man (Vav), the same one who is the intermediate point between light and darkness and who will be the one who will come to the arranged place. It means that the objective of man was also already assumed and it was not simply to populate the Earth but to make a distinction between light and darkness, being helped by superior beings who are at the service of God. That is why the Hei surrounds him, since he would not be alone and also the added destiny of coming to this reality to LEARN, knowing and experiencing development in this visible, superfluous and vulnerable existence would be established for him.

Based on another method, continuing with these means of investigation that can take a long time, the first word of the Bible: "barashit" calculates 913, which corresponds to 900 + 10 + 3. The 900 usually starts from the final Tzadik, but the sofít (final) letters are rarely used in numerology, so their equivalence must come from other components, such as the case of R (200) + SH (300) + T (400), which are precisely the 3 endings of the alphabet Thus we have R.SH.T. Along with Y/I (10) plus G (3). We have already seen the equivalences of R and T, but not that of Shin (fire), Yod (excellence) and Gimel (advancement). In this way, what is usually translated into Spanish as "in the beginning", more profoundly, would refer to more than one thing: in the elementary sense it speaks, once again, of a place that would be cemented to begin a divine work, in which the consciousness of God would act and there would be information brought by the hand of divinity for a specific job, which would leave a remarkable record (BRASHIT). However, under the secondary system, the numerical one (which gives 913), would refer to the beginning of the action of consciousness and knowledge, leaving a mark and acting by divine will in this reality; maintaining this work and effort until the work is finished. Although, Barashit is also composed of two groups:

Bará (create) and shit (settle). This creation that was established also forms two numerical groups: 203 (Bará) and 710 (Shit), thus referring to RG and SHTI, which returns to the same thing, but said in a different way.

Assuming that before man intervened, light came to reject darkness, it is relevant to take into account the words cited, since each one implies a broad concept. Therefore, it is possible that Tehom (abyss) that is under the "waters of the sea" -which Enoch described-, are something that is not subject only to the world of matter, or rather it came from the invisible to the visible –maybe simply highlighted by the fact that before, since there was no light, what had been up to now was not observable. Also because it can have forms, entities, energies and forces of various frequencies or vibrations. Now, we find that due to Sofia's desire to create a being without the approval of the Father, and since she had great power in her, she not only created Sakla but "something" in aggressive activity, enveloping and intermingled with flame and darkness. This aspect was seen or changed to a water so dark that, according to what is used, it would be like pure oil (this crude, and the lava, it would not be strange that they were residual memories of that event), while the purified part would be as translucent as the Diamond. With this process, this potential was separated from its elements, creating a huge hole under itself and enveloping it in those waters. Immediately after, from the black part taken to the bottom, located in the abysmal of the emerging ocean, a cluster of substances solidified that came to the surface, revealing the dry land, full of all kinds of minerals. The maritime zone and the dry zone would be arranged on a strong platform with beams, seen from afar as a Swiss cheese, and inside the world would be a dark hole, the abyss. Here would remain 1/8 of the cleaned aquifer mass (not in proportion but according to the division that was made into factions), while with the other 7 parts Mercury, Venus, Mars,

Jupiter, Saturn, Uranus and Neptune were apparently built. This theory does not explain where the Kuiper belt, the Trojan belt (between Mars and Jupiter) or the Oort cloud came from, nor how the Moon and the Sun emerged, how Pluto was added, or other mythological aspects such as the planet Nibiru, but it is, for the moment, what the sources refer to (the other may be a later scenario that is somehow exposed at a certain moment).

Genesis quotes: "*And Aretz was in chaos and desolation, and [darkness] was on the face of the Abyss, and the spirit of Elohim hovered over the face of the waters.*" (Genesis 1:2, 1960 Reina Valera version) What Elohim, or what god was this? It was clearly Sofia. It does not say "Iehovah Elohim", but only Elohim, a plural way of singularly describing the deity, which, in this case, is not personified but acts as a "wind" (wind and spirit are the same under the Hebrew language and Greek). Sometimes it describes an individual and other times a group. So, at the beginning of what concerns us as earthlings, the Father created the biblical heavens -that is, those referred to in the history of Israel-, as well as the spiritual universe, and then he created the entire material world and, with it, the purple physics we know as "Earth". Let us reiterate that in the entire event of said genetic creation, the origin of darkness or the abyss is not explained, and yet it only clarifies that light and darkness were distinguished or differentiated, but nothing was defined around the Abyss, so in fact, he has had to investigate step by step to find the true nature and identity of that light that took place at the beginning. This illumination is soon understood to be Sofia intervening in what is happening, having the approval of the Most High, the same one who approves that Jehovah represents him, thus showing that all things are always attributed to Jehovah, as the Father's delegate on this plane.

Returning for a moment to the manuscripts in the Nag Hammadi Library, let us take a look at a Coptic text that we have already used: "*But when the Faith gained [understanding about this] insolence of [the] ruling [of Sakla], it sent its breath, she [forced him] and plunged him into [the] Tartarus. [Today] there, heaven was strengthened with its earth by Wisdom [and] Yaldabaoth, who is below all.*" (Coptic Creation Apocryphon 102:32 to 103:2) This First Ruler believed that everything was at his disposal and created everything he could think of without knowing where he came from or the unlimited power he possessed, but Sophia ridiculed him. and also threw him into Tartarus (he did not create the world, but he did create the aspects contrary to the pure). If this material really says what it seems to say, we're looking at a direct reference to the Beast being sent down into the very chaos from which it emerged. Tartarus, according to Greek literature, is a place in the underworld; the Bible defines it as a prison for rebellious angels, beyond Hades, even deeper. This seemed to be the fate of Sakla, who is possibly the same one who will soon appear. Sofia estimated this about him because of his daring: «*When [the] Arcane [saw himself] in his own greatness - and he is the one who lives alone – [but] he saw nothing, except water and darkness. Then he thought that [he] was the only existence. [And] his pen [was] made by the word that is expressed in a spirit back and forth in the water. And when that spirit appeared [to] the Arcanum [the] watery content was separated on one side, and what was [was] dry was separated from the other. And from the field, he created a home and named it 'heaven'. Also from the field, the Arcane [created] a ladder and called [it] 'earth'.*» (Apocryphon of Coptic Creation 100:30 to 101:10) Thinking about this, let us leave on the table of possibilities the question that light -the glorious manifestation of Sophia on the waters- came to quell the power of chaos, separating the substances from each other and generating a vacuum under said aqueous mass, which would be seen as an enormous dark cliff, but not only in material matters but in relation to entities adjacent to it.

A text from one of the two sons of Joseph, son of Israel, quotes: «*You who made heaven and earth with all their end; who [has] chained the sea by his word of command, which borders the Abyss and [has] sealed with his terrible and glorious name; all these things which tremble and tremble at his power* ..." (Prayer of Manasseh 1:2-4) Moses wrote that the "Ruach" of the deity "fluttered", thus being a way of speaking of the nerves of Wisdom in the face of the chaos that blossomed: «*Although in the darkness she had forgotten her ignorance, she began to feel ashamed and agitated. This agitation is the movement from one side to the other. [...] She ascended not to her own eternal realm, but, instead, to a position just above her son. She would remain in this Ninth Heaven until she restored what was missing. A voice called from the exalted heavenly realm: Humanity exists, and the Offspring of Humanity! The first ruler, Jaldabaoz, heard the voice and thought it had come from his Mother. He did not realize its origin. It came from his holy Father, the perfectly complete Forethought, the image of the Invisible, that is, the father of all, through whom all was born, the first Humanity. She taught these things, and revealed herself in human form. The entire realm of the first ruler trembled and the foundations of hell shook. The bottom of the waters above the material world was illuminated by the image that had appeared. When all the authorities and the first ruler looked at this apparition, they saw the whole background because it was illuminated. And through the light they saw the shape of the image in the water.*" (Secret Book of John 8:4-11)

So far we have seen how Moses' account of origins, quoted in the opening chapter of Genesis, corresponds with the cosmogonic claims of many other cultures – if not virtually all. After seeing how this material creation appeared, we have to question the order of things, since science has its own theories regarding the formation of the universe. Although, Moses must have taken his sources from an earlier root, such as Enoch or writings that, as sheltered by the

Egyptian monarchy, he would have available, in the most powerful and cultured kingdom of that time, or simply the angel that accompanied him advised him above all. It could also be the celestial messenger who was with Moses all the time who dictated these mysteries to him. On the other hand, Enoch clearly predates Moses, and in his travels he had the great opportunity and privilege of speaking with the Ancient of Days, who would tell him in great detail what Aaron's younger brother, Moses, would write millennia later. God speaks to Enoch and tells him that while his creation is at rest, he is not. Although it is strange that God gets tired, being God, and that Enoch could see him without having been "pulverized", his account tells us that He brought the visible from the invisible, so the invisible previously existed. For his part, Ezra reported the following: «*I told him: "Sir, you have spoken at the beginning of Creation, and you said on the First Day, 'That heaven and earth were made', and your word [was] the work accomplished. Then the spirit blew, and darkness and silence [had] embraced all, the sound of the human voice was not yet there. Then he sent a ray of light to become manifest from his habitation, so that his work might be seen.*» (4th Ezra 6:38-40) Evidently that "spirit" that "blew" was the same as the Ruach that revealed the light that Ezra expresses as "lightning".

We have noted above the teaching of the Ancient of Days made to Enoch. Although this extension of the book of Enoch is rare, since it comes from Greek and Slovak texts, it is important to look for it and investigate it. The translation that I will expose I have made by comparing and slowly going from the characters of the Hebrew Book of Enoch of the Library of Jerusalem. There are big differences with the popular edition, such as that in the official version it speaks of "Adoil", while in Hebrew it refers to Uriel, which means: "Light of God". Whether it is the angel Uriel himself, I do not know, but it does not appear to be, to my judgment.

Then, the matter of "giving birth" is not textual but is used from the usual phrase to refer to a birth. When this takes place, God was in the midst of the light that appeared from Uriel. Then he says that the light "irjef", that is, "fluttered", the same as Genesis 1, when it speaks of the Ruach Elohim. It also uses the definition of "mareh", in association with a "vision" or "likeness", although the word comes from the root "mera", which is "rebellion". Then it is understood that, just manifesting this first "olam" (world), God makes a throne above said original Creation and then invites the Light (Sophia) to be with him, being the base of "elevated" things." (from the Hebrew "elyonim"). Finally he ascends to his throne, it seems, to declare the second part of Creation that is emerging according to his will. I reiterate that the word "olam" defines "world", but it encompasses a creation, universe, cosmos or the entire set of created things. Analyzing Enoch's words, we see that the order for the matter to manifest came from Above towards what was "below", I suppose that that which underlies the sky, below the superior realities. Hence the talk of "the very lowest parts", or the deepest, as the idea of the Abyss. The "womb of great light" that Enoch writes about could be what contained all the potential of the visible cosmos, a matter that explains how it turned out to be "waste", leaving the invisible to be visible. The unknown in this section is, why was he invisible? Why was there no light? Why duplicate what already exists but in the physical dimension? Or, why was it already thought in the bosom of God, even more so it was not embodied in material "reality"? If God was in the midst of that great light that originated, where did that great light end up? Was it the Big Bang? Or was it just the light around him?

He says that in that creation light arises from light, the initial potential of the universe being the light from "Above". In this way, a great age or "day" would begin, before the one taught by Moses. This is how that "great existence" appeared, which God had in mind, so that fact, having it in his creative thought, was what, possibly, made all that accumulation, a majesty that was initially invisible, and then visible, when it manifested itself in our reality. What follows again draws attention, as God orders the light to return to its point of origin, on high, and also tells it to "repair itself", apparently locating itself in a place next to the God's own throne. We have to confess that this description corresponds to Sophia (Wisdom), the material Holy Spirit, the light who appeared at first to remove the deficiency, and later ascended to the throne side of God to be with Him. Furthermore, he adds that the Light it was to be a "foundation" for "higher things," being the highest, above which "there is nothing else." Being a "foundation" is nothing more than coming to become, be part of or direct a work of preparing the foundations and structural bases of something. In this case, of "elevated things", which come to be identified with what is in the divine dwelling, not necessarily what is on the veil that separates the superior from the inferior. With which, that Light had to take care of the "foundations" that would shape the image of the things "above" below. Next, the matter that above the light there is nothing else, must refer to that entity identified as Light, above which there is no more "possible" reality, "she" being the limit of reality, which the texts of Nag Hammadi mention as "the veil", an important point, also due to its symbolism the evening of Jesus' crucifixion, when the "veil" of the Temple was torn from top to bottom. Or it can also represent the location of God, on which nothing exists, since it is the maximum of his creation.

As we have seen previously, Enoch's words continue, stating that God said: «*And I called second of the low [things], and I said: Let out of the invisible things spilled visible and Arjas come out hard and heavy and very black. And I said: open up, Arjas, and be born of you and be visible. And he released and came out [the] world of darkness very great, [from among] what was made of all the previous creations. And I saw that it was good, and I said to him: move [down] down and settle, and be the foundation of low [things]. And there is nothing else under the darkness, and it advanced and came out, and foundation of low [things].*" (2 Enoch 26:1-4) Although this corresponds with the Hebrew text chapter YA:YD (11:14), I follow the numbered pattern of the non-Hebrew version. Now, once the first was finished, he located himself and commanded the second, referring to it as the "tajtoním" which is plural of "rear" or "edge", although it is translated as "low" or "low place". So he says that the "hard" came out, that is, the material could be the initial Light, and now, the second, tangible "mass", so it is hard and heavy. The question of it being "very black" –different from the other translations, which say: "very red" (as can be seen above, in the textual description that I made before)- is rare. In this case, that heavy, hard and black mass decomposed or defragmented, giving rise to a whole concrete form, defined as "Olam-Joshej" ("creation of darkness" or "world of darkness"). If we look closely, we see that two realities have already originated. Both from nothing or "invisible" and both from an apparent "explosion" or fragmentation, defined as "delivery" or "opening". That is why he speaks again of the appearance of an "olam" (world, cosmos, universe, eternity). However, while before he spoke of light, now he speaks of darkness -almost always in the singular. Although, before, he referred to the fact that there is nothing above the light, despite the fact that here he mentions that in the second step, this part defined the limitation of darkness, below which there is nothing. What can that mean? I could say, for example, that there is nothing brighter than light and there is nothing blacker than darkness. What this may suggest, too, is that the first creation, that emerging from Uriel, is that of the Kingdom of Light, and the subsequent one, that of Arjas, is that of the Material Kingdom. Moreover, that Hebrew word, "Arjas" (ARCJ.S.), could give rise to the Greek "Argí", used by John in his Gospel (Cap. 1:1), and in which was the Logos, as principle of formation and potential of the whole, housed in the bosom of the Father. No wonder, since the Septuagint translated "Barashit" into Greek as "In Argi" (in [the] Beginning). In Aramaic "arjah" is "prolongation " (see Daniel 7:12), not necessarily a beginning with no aspect prior to it.

Now, the "age", which Enoch defines as "dark" -literally "darkness" or "dark"-, seems to be the opposite polarity to the Kingdom of Light. Essentially the universe we observe, where the majority is virtually black, and obviously "very great." The strange thing is that Arjas too, is the beginning of creation –or whatever- did something similar to Light, Uriel, but going below what is in the lower part, becoming firm. It is as if he, for his part, should be the essence, base or foundation for all this reality that is mostly dark, I suppose, outer space. This would explain that, unlike the Kingdom of Light and the Material Kingdom, there are no more real creations or of this nature that encompasses all of reality. What follows is already taking the form of something that we have more or less heard about: «*And I ordered that there they should be taken from light and darkness, and I said: "Be thick and surrounded by light", and I separated it and [it] became water. And I separated it over the darkness under the light and so I separated the waters and there was an abyss, and I hardened it into light, a circle of water, and I created 7 circles in the middle, and solid as a wet and dry cup, and the glass and ice of the circle of the waters and [the] rest of the elements; and I showed each of them their way, to the 7 stars, to each of them in their rakiah, and they all went [like this]. And I saw that this was good, and I differentiated between the light and the darkness, that is, in the middle of the waters, between these and between these, and I said to the light, it is the day, and to the darkness, I said that it was night; and it was late and [the] other was morning, and this is the First Day.*" (2 Enoch 27:1-2) Once again

I quote this chapter, but showing the distinction of translations, considering the importance of history. So it seems that what comes to form from light and darkness was "water", a giant mass of water -in the plural: "waters". But what would it mean to become "thick" and "surrounded by light"? The "thick" comes from the Hebrew "ebah", which can be associated, in a certain sense, with something solid. The "surrounded" does not refer to being protected by a field of light, but to something "round", identifying a circumference (clearly the ancients knew that the Earth was round, only that the millennia plunged people into ignorance and ambiguous legends), which is why in Hebrew it speaks of "misebab" ("sibub" is "turn"). Likewise, we see here that after this mass of water appears, because of the light and the darkness – be careful, it does not say "darkness", and they are not the same – the darkness is separated. Within itself a deliberate split is generated that takes place "mitajat la-or" ("below the light", "at the end of the light" or "beyond the light").

Also, when referring to that solidity and circumference of light, he speaks of "araká" (being its root: "rek", that is: "empty"), that is, "be emptied" or "empty" and there were "waters". This doesn't seem to make sense. Chemically, only when the "gas" "solidifies" does it compose water. In any case, from "rak" or "rek" the verbal form of "separate" or "differentiate" is taken, making a "space" between them. Space is precisely an apparent void, and that suggests having been born in the water, causing an "abyss" or "hole" to appear (it is confusing to differentiate in these definitions and stories between the darkness of chaos in the abyss and the darkness that encompasses outer space). If we remove the action of gravity and mold an aquifer in space, with the necessary means, separating the mass of water would leave, between one quantity of H 2 O and another, a vacuum, precisely. The issue is the dimensionality and location of that "empty", because when speaking of "abyss" it resembles a fall, precipitation or immense depth. So, he goes on to say that with the force that was in the light he made the water take on a circular shape - which some would think refers to power, energy, electricity or other technology from the initial mind. But not only one shape but 7, in addition to the base (That is, there would be 8 spheres). Those masses of water were "ba-tebaj" (in the middle), being curious that, when speaking of the Earth, he established those other spheres "in the middle", but, in the middle of what? In the middle of the light? Well, the following makes it clear that he is talking about the planets of the solar system, when talking about their visual form, "similar to a wet and dry cup", but solid, and the water itself would be related to crystal and ice. This is how Uranus and Neptune are still seen in photographs today, but, if this were the case, 7 large bodies had started out being similar to the perception of the eye, as round-shaped liquid masses, which would also have the rest of their "isodots". (fundamentals). The word "isodot" is also translated as "bases", "components" or "elements", implying that they were not only aquifers but also had other basic substrates in their total composition. According to Enoch, the Most High would have predetermined their guidelines and "his sky", which we can deduce that synthesizes the parameters of our 7 neighboring stars, although, added to physical laws such as translation and rotation, or their work as a body and, definitely, directed by elemental forces and energies, described as "kojabím" (stars) – that the nations would have idolized as "gods".

There are also 7 stars or Spirits of God, and likewise 7 elemental skies. So, we can suggest that the 7 orbits of these planets identify the 7 heavens of this solar system, and are ruled by 7 archangels. All this, of course, in comparison with the 7 heavenly kingdoms – unless they were the same. By doing this, it would be possible to remit the strength and authority that Sakla would have given to the 7 that he delegated on the "7 spheres of heaven". That would have been the famous parting of the waters: «"*Once more, on the second day, he created the spirit of expansion, and commanded to divide and separate the waters, so that one part can move upwards and the other. another will remain low."* (4th Ezra 6:41) This is how they isolated from the mass that appeared and that they did not want it to keep its essence, so they put it under the waters – it seems, only from the Earth. There is also the classic part that is call one and the other with specific names: Light (Day) and Darkness (Night), but, according to what you read, they are also described as "afternoon" and "morning", respectively – although the order seems to be reversed. one afternoon and the other tomorrow came the first era or age. If light is Day, how is it that evening and morning are also "day"? This account of Enoch, unlike that of Moses, begins by speaking of other formations, for so it is possible that the beginning written by Moses referred precisely to what we have just reviewed, but summarized, or assuming that the story was already known Ironically in theism it is thought that the Creator made "the heavens" and "the Earth" as an inaugural step in the embodiment of his thought in this reality, but that Heaven is never defined, and the Earth already appears to be a complete disorder. Let us remember that in Hebrew it says: " *Barashit bará Elohim et ha-Shamaim be et ha-Artz* ", which could be translated as: " *head created [the] deity those Heavens and that Earth.*" You have to see the similarity between "barashit" and "bará", since the second is the root of the first, which is why it cannot say "in the beginning", otherwise that phrase would be: "in the beginning it began". It is not correct. On the other hand, "barashit" refers directly to "creation", since it comes from "bará" (created). For example, in the book of the prophet Jeremiah (chapters 26:1, 27:1, 28:1 and 49:34) he speaks of the "beginning" or "beginning" of certain reigns of the monarchy of Judah (specifically Josiah and Zedekiah), under the word "barashit". Now, the beginning or beginning speaks of preeminent things: "heaven and earth."

The Heavens

In English, that heaven described there is called "heaven" (that of religions or spiritual concepts), that is, God's heaven, it is not from the atmosphere (sky), and assuming that this is the type of heaven to which it refers, in the book of the prophet Isaiah (cap. 65:17 and 66:22) and according to the 2nd letter of Peter (cap. 3:13) it is maintained that there will be "*new heavens and a new Earth.*" So where will we end up? Conversely, Baruch wrote: "*For he knew that his time is short, but that heaven and earth endure forever.*" (2nd Baruch 19:2) So, if we do not go to heaven, as various monotheistic tendencies erroneously teach, nor will we inherit the Earth –contrary to what Isaiah recalls-, where will we be? Suspended in space? It is necessary to see that the Earth is located in the middle of the cosmos and is an orb among hundreds of billions -or more- that statistically seem to exist in the entire universe (being that it is also estimated that the Earth is relatively recent compared to other stellar bodies). If the first thing that Elohim created was the spiritual or etheric sky, where was the Earth stationed? It is strange that galaxies or the Big Bang are not mentioned in this part of the Biblical account, or even something similar – or yes, but it has not been understood. One could suppose that the narration does not speak of the beginning of the cosmos but of our planet and of a "temporary" heaven, since if God lives there, why should he renew his dwelling? Is your home bad? In the event that it was a seasonal place from which one works on Earth, then it is obvious that while there is a problem on this globe, it is also necessary to have a momentary "heaven" to act on it.

One of the common heavens, regularly referred to in Hebrew scripture, appears to be an establishment on Earth from which work is done for humanity. That sky, like others mentioned, is assumed to have something to do with a planetary orbit and its sphere. We can wield something of this caliber from the verses of old Thoth: « *There are two regions between this life and the Great One, traversed by the Souls that depart from this Earth; Duat, the home of the powers of illusion; Sekhet Hetspet, the House of the Gods. Osiris, the symbol of the guardian of the portal, who returns the souls of unworthy men. Beyond lies the sphere of the heaven-born powers, Arulu, the land where the Great Ones have passed. There, when my work among men is done, I will join the Great Ones of my Ancient home."* (Table 14. Emerald Tablets of Thoth) This "Arulu", located very far from our planet, shares an amazing phonetic and etymological resemblance to the Sumerian Alalu, who, if we remember, was the ruler of the divine abode before he was dethroned. by Anu. Let us consider that when we speak of the primordial Earth, it is in deplorable or damaged conditions while the sky is born clean. Moreover, it creates the sky, but they do not specify it, with which it is presumed that it came out elegantly with only the creative thought of God. On the other hand, with our globe this does not seem to be the case, to such a degree that, instead of forming quickly, as it seems to have been achieved with the great sky, with this tiny planet it took the deity 6 whole days to organize it. It should be noted that this is highly suspicious, adding to the fact that Heaven being a healthy site is considered to be renewed or reinstated. The Biblical references to Moses do not seem to refer to the making of heaven - there is no further explanatory development of how it was made - nor to the entire planet as such, since its consistency already seemed to be glimpsed, at least to a certain degree.

The Sumerian texts suggest that the sky that corresponded to the abode of the gods come from above was the "house of An/Anu", or the "celestial abode". They would have reported that this world was apparently uninhabited when they arrived almost half a million years ago, and the first thing they carried out was the construction of their house-home that came from the confines of space, the house of heaven, and then they leveled and they decorated the land that was later called Eridu – arduous work carried out by one of the sons of An, called Ea in Sumerian and Enki in Akkadian. Hence, these prehistoric beings are attributed to being part of the deity, in some sense, or simply to have come to be used in order to achieve the divine objective: "... you established the earth, *and it stands firm* ." (Psalm 119:90) David's words would not confuse "founding" the foundations of the Earth with the fact of preparing a suitable land for the construction of a city, with which David would speak of the beginning of our space house, long before of the arrival of Ea/Enki to establish the primeval civilization. The Israelite king knew well that the merits of terrestrial organization and its modification fell into the hands of the Hebrew God: "*He shakes the earth from its place and trembles its columns.*" (Job 9:6) Any skeptic would say that these may be misrepresentations that would appear over time or mere flukes, but the scientific principle of Occam's Razor holds that when you seek to understand a phenomenon, which can be done with few assumptions or concepts it is done in vain with more, or in other words, when faced with a question, the most obvious or simple answer is usually the correct one.

Moving on to the next point in the story told by Moses, we find that it says: «*And God called the light Day, and the darkness he called Night. And the evening and the morning were one day.*" (Genesis 1:5) Although this literary work maintains that "in the beginning Elohim created the heavens and the Earth", this introduction would be no more than that: the opening of the story. And that is so since the structure of Heaven does not really come to be considered until several stages have already passed. Consequently, the universe must already exist in some way, and what is assumed to be relevant, considering what concerns us, is to tell us the version of our planet. Now, in relation to light and darkness, specific names are given to both polarities: light is Day and darkness is Night. In modern terms we understand that if we talk about a day, only, it indicates 24 hours. However, when we also refer to the night, we assume that that day includes 12 hours of light and 12 hours of darkness (technically, and starting from the Equator), as it used to be applied in the past. The difference between night and day is found in the time it takes for the Earth to rotate on itself (rotation), since it constantly receives the direct rays of the sun. So, the first thing would be a long and sinister night in the origins of this world, which would be indefinite until light appeared and we identified it as day –since "afternoon" and "morning" would not be ways of seeing the sky, as orange, because the sun has not yet appeared, since the sun king is still manifest, this pattern is followed. It is elementary to conceive in the mind that this luminous appearance was the sun, since to think otherwise would be to accept that said light was something symbolic or representative, alluding to another aspect, not of a 12 or 24 hour day – not even of a matter of time. time, necessarily, or at least as we now measure it. I highlight this because there are still "three days" left for the official appearance of the sun, and without the sun, what day would there be? Although, it does not seem to speak of the days as we understand them but of something "spiritual" and something "symbolic". So that group of initial things, which take place at the beginning, are conceived as "Yom Echad" (day one), with the objective of reducing the priority of darkness. In other words, the deity was bothered by the presence and apparent supremacy of the night and for that reason he made the light manifest and set the standard to know how to define between one and the other – an aspect that he would assume as allegorical and metaphorical, since from being literal sounds ridiculous: light and dark are intrinsically incompatible.

In the case of the book of Jubilee, in relation to the separation process already alluded to, Moses wrote: «*And on the second day [it was on] that he created the Rakiah in the midst of the waters, and the waters were divided on that day - half of them went over and half of them went down below the Rakiah (which was) in the middle over the entire face of the earth. And this was the only work [that] (God), [has] created on the second day*." (Jubilee 2:4) Another scribe already wrote down, in a generic way: «*In his word the stars were fixed in their places, and he knows the number of the stars. He seeks the deep and its treasures, but has measured the sea and its contents; who has bounded the sea in the midst of the waters, and by his word he has suspended the earth above the water. The sky has stretched out like a dome and secured it over the waters ...*» (4th Ezra 16:56-59) By divine order the stars follow a course and are in an exact position, being known in number. If it is a matter of finding the abyss and what is in it, one will find that it is located under the oceans, being perfectly located and delimited in relation to the seas, and the seas in relation to themselves. The guidelines of science are flawed here, considering that the Earth floats on water tanks and the other parts have the celestial vault as their shell, like the glass of what we know today as a safety helmet. It is ironic that it seems so simple to understand and that it was necessary to go around multiple references, but that is what the phrase of the Benjaminite to the congregation in Thessalonica consists of: «*Examine everything; hold fast the good*» (1 Thessalonians 5:21)

"*And God said, Let there be expansion in the midst of the waters, and let it divide the waters from the waters.*" (Genesis 1:6) Continuing with the base of the RVA 60 translation (version made by Casidoro de Reina and Cipriano de Valera, taken in the year 1960), we have that Elohim gave a second order, in this case to separate the waters. We then continue to see that there are apparently obvious difficulties in its creation, since it is forced to decree a separation between what it decrees is "good", next to what is obviously not. What now had to be seen was a "differentiation" named "rakiyah", whose root in the Hebrew language is "rek", which means "empty". It is important to note that there are two common words to call the heavenly place: "shamayim" and "rakiyot" (in the singular they are "shamai" and "rakiyá", respectively). Both are translated as "heavens", but they are not the same. The concept of "shamayim" is also Akkadian-Sumerian and identified the celestial abode, meaning "rocket people", coming from various explanations, either under the Niburian, Akkadian, Aramaic or Hebrew prism. For example: "sham" and "maim" ("there" + "waters"); "shama" and "im" ("over there" + "masses"); "shem" and "aim" ("rocket" + "peoples"), and a few other forms that derive from the same thing. The word "shem" has its own history, since the Akkadians used it to refer to the rockets that the gods used to go out and enter the Earth, starting from their shuttle (on whose bases the post-diluvians tried to relocate it: the Babel Tower). Moreover, this definition (shem) passed into Semitic as the equivalent of "name", in line with the previous meaning: "fate" and "launch". So that place above was where the people of destiny were, the people of the rocket, the people from above, not referring to "shem" as a name to give to several places but to define only one. For its part, Rakiyá or Rakiá denotes more than one site, being a generic way of referring to a place, location, room, space, dimension or defined, limited or specific region. For example, the Hebrew texts speak of 7 heavens and 10 heavens, and Paul referred to a 3rd heaven; all these translated from the Hebrew "rakiyah". On the other hand, the Heaven created in Genesis and that will be replaced in the new age, is defined as shamaim. Therefore, removing Barashit I:VI from the Hebrew version, we find a rakiyah that is established between waters and waters to distinguish between them.

When Elohim said that there was Rakiyá in the middle of the waters, he exposed that a specific space was decreed to divide the masses of water. The typical translations pass the voice rakiyá to "expansion", "vault", "firmament" or even "horizon", being also passed to "heaven" or "heavens". What it literally says here is that it creates a void or unspecified dimension, which puts one mass of water in one place and the other in another, and in between there is nothing of that – that is, of some water or of the other. We then have what follows: «*And God made the expanse, and separated the waters that were under the expanse, from the waters that were above the expanse. And it was like that.*" (Genesis 1:7) So we see here that this "void" appeared between said aquatic mass and placed a part above and another part below that "hole" or "space". Common words are not used in this vocabulary, such as expansion ("harjabá"), vault ("gag mekumar", "kimron" or "kuj" or "gumcha", in Hebrew, although the definition itself is not mentioned in the Bible. in Spanish). For its part, "rakiyá" is still understood as firmament in modern Hebrew, but by mere custom and/or association, not because it respects the etymological roots of the word itself. If this fact is acceptable, it could be respected that the terrestrial atmosphere was formed, but that does not respond to the implication of celestial forces intervening in it, unless they were there and we do not see them. Baruch, exalting the Lord, said: «*O, [Lord] that you have made the earth, you listen to me, that you have fixed the firmament by the word, and you have made firm the height of heaven by the spirit, which you have asked from the beginning of the world that did not yet exist, and they obey you; that you have sent through the air of your wink, and I have seen the things that are to be like those things that you are doing. May the article be with great thought from the hosts before you. Also the innumerable holy beings, which [you] made for yourself from the beginning, of flame and fire, who stand around your throne in indignation towards the command.*" (2nd Baruch 21:4-6)

Then the narrative is added: «*And God called the expanse Heaven. And the evening and the morning were the second day.*" (Genesis 1:8) So that emptiness receives the name of "heavens", that is, that rakiya receives the name of "shamayim", a special accumulation of curiosities, especially due to the symbolism and meaning of both words. Thus, he explains that this space or void has been defined since then as the abode of the celestial forces, being observable as a tenuous crust that encompasses the ozone, or even the ozone itself. A noticeable difference is that the initial Heaven is not specified in location, but this one that now appears is as if it were not very distant, as if it were the atmosphere that we have. In other words, the first thing to be created is said to be the Shamaim and then the Aretz, but later on it says that when the Earth was already there, the sky was created as an intermediate between the waters, which places the probability that the first heaven Mentioned was the Kingdom of the Father, the abode of light, and the other, mentioned correlatively, came to be our atmosphere, not exempt, of course, from the prescribed and so relevant phenomena that are added. The Heaven that is spoken of as the opening of the book of Moses, is already considered "Shamaim", just as it is, but the one that separates Day One from Day Two began as a void or space that was later defined as Shamaim, assuming the place of the highest Shamaim to be the zone from where the messengers of God would act in relation to the Earth. For this reason, the High Sky will not disappear, but rather the functions for which this terrestrial Heaven was established, not in its quality of atmosphere but with respect to the spiritual work that will no longer be necessary for the future.

Returning to the second book of Enoch –sometimes identified as the 3rd Enoch or Apocalypse of Enoch (it depends on whether it is 1st, 2nd or 3rd, nothing more than the individuals who obtain the scrolls or fragments and how they organize them)-, we see that the The Most High continues to narrate things to the prophet as they were in the beginning, and he also cites what Moses describes in the stage that receives the name of Yom Sheni (Second Day): «Thus I made firm the circle of the heavens, and I said that *the low waters that were under Heaven in a whole and dried up their heaps, and it was so. And from among the piles I created large hard rocks. And from between the rocks I caught dryness, and it happened that the earth dried up. And for in the middle of the pure Earth depth happened and there was an abyss. That one, the sea, I stored in one place and tied them by the sand. And I said to the sea: look, I put a perpetual limit on you, and you [must] not fracture yourself. Thus I made the rakiah and its foundation above, from [among] the waters. That day is the head of the creation that I created myself, and it was afternoon and another morning, it was the Second Day.*» (2nd Enoch 28:1-4) On another previous occasion, he wrote: "*I saw how the winds spread the veil of Heaven on high and how they have their place between Heaven and earth: they are the pillars of Heaven ...*" (1st Enoch 18:3) According to what is read from this personal experience of the prophet, those "winds" or "spirits" (ruach, both in Hebrew) as a whole are those that determine the dimensions of the atmosphere and its volume and pressure come to be its support.

On the other hand, dimensional portals have been studied, openings that lead to other dimensions, which seem to exist in specific points on Earth -and it would not be surprising if they also existed in other celestial abodes- similar to the stellar bridges that are said to exist. in ufology that exist throughout the universe, which, upon entering them, vehicles are launched to supernatural distances. There are those who find ideas put forth in this regard in the Aenoic texts, perhaps not as dimensional portals but as accesses that are not visible or feasible for the common man: "*I saw the treasures of the wind-spirits and I saw that with them He (God) has adorned all creation and the foundations of the Earth; and I also saw the cornerstone of the Earth and the four winds that uphold the Earth and the firmament; I saw how the winds extend the veil of heaven on high and how they have their place between Heaven and Earth: they are the columns of heaven; I saw the winds that spin and drive through the orbits of the sun and the stars in their rooms; I saw the spirit-winds that hold the clouds above the earth; I saw the ways of angels; I saw in the ends of the Earth the firmament on high. Then I went south and saw a place that burned day and night, where there were seven mountains of precious stones, three on the eastern side and three on the south side. [...] I saw a burning fire, and beyond those mountains is a region where the great earth ends, and the heavens culminate there. Then I was shown a deep abyss between columns of celestial fire, and I saw in it columns of fire that descended to the bottom and whose height and depth were immeasurable; and beyond this abyss I saw a place over which the firmament did not extend, under which there were no foundations of the earth either; on which there was neither water nor birds, but was a desert and terrible place. There I saw seven stars (watchers) similar to great mountains, which were burning, and when I asked about this, the angel told me: 'This place is the end of heaven and earth; it has become the prison of the stars and of the powers of heaven... "* (1 Enoch 18:1-15)

Enoch says that with the winds or spirits God has " *adorned creation* ", but is it an invisible decorum? If he says that he saw the winds that make the sun go round, it is absurd for him to say it literally, since in space it is assumed that there is no wind, so he seems to be referring to "forces" or "spirits" that direct the sun. movement of things and make the laws they obey come true: « *Those who do not sleep bless you; they are before your Glory and bless, praise and extol saying: "Holy, Holy, holy is the Lord of spirits, He fills the earth with spirits."* (1st Enoch 39:12) The above quote explains many things, such as, for example, which are the columns that support Heaven, and of which the Bible speaks so much: «O who among all humans can contemplate all the *works of the Heavens or the angular columns on which they rest? And who sees a soul or a spirit and can come back to tell it? Or go up and see all its confines and think or act like them?* » (1st Enoch 93:12) It is as if it were referring to some force or power, visible in some way, that controls or holds our atmospheric and/or orbital sky. It makes it seem, at first glance, something similar to a majestic energy or spirit -or a group of them controlling 4 strategic points- that maintains the celestial mechanism of this first heaven, supporting it on 4 bases, which can also correspond to the cardinal points. To this we would add another reference from Enoch who says, next: « *Who among all humans can know what is the length of the Heavens and what is their height or how they are supported or how great is the number of the stars, and where they rest? all your lights* » (1st Enoch 93:14) When speaking of the stars, it covers the number of astral bodies that we observe, which comprise an enormous amount and are found from Alpha Centauri to a certain distance before the Andromeda galaxy – which is already the most distant, no matter how large they are, they are not seen with the naked eye. The stars in relation to us are neighbors of local systems within the Milky Way, since it is their ensemble that determines the majestic vision of an entire galaxy – since the most titanic visions in space are really immeasurable regions of deposits of different types. of gases.

Where was Enoch? He wasn't asleep or hallucinating, but the descriptions he gives don't seem to fit anything we know or understand from a quick read. Would he have been transported to another dimension or to another planet? Not necessarily, maybe he just looked at our world with different eyes. But what would he mean by "the end of Heaven and Earth"? All the first thing Enoch observes, where can it be located? Where on Earth? If you are on the surface, where did you see that "cornerstone"? What winds or spirits are those that sustain the Earth? Furthermore, how did you see the columns of the Earth? We can intuit that he was not on the surface but below it, although he affirms that he saw the columns of heaven, what columns are these and where are they? This man seems to be talking about a trip in a flying saucer, with which the stratosphere is shown, he is lowered into the abysses and so on. Otherwise he could not have spoken, or even reached, a region where "the firmament does not extend." Was it a cave? But he says there were no "foundations of the Earth" either. Enoch's experiences are fabulous and impressive, but they make it clear that everything is moved by forces that science or has taught us exist, not in that way. Enoch further added: «*From there I went to the ends of the earth and saw there great beasts different from each other and also birds that differed in appearance, beauty and trills. To the east of those beasts I saw the end of the Earth, where Heaven rests, and where the portals of Heaven open. I saw how the stars of the heavens are born and the portals from which they come and I noted the exits of each one of the stars, according to their number, name, course and position and according to their time and months [of delegation], as he showed them to me. Uriel, one of the Watchers. And he showed me and wrote for me everything, he even wrote for me their names according to their times.*» (1st Enoch 33:1-4) If he was in the hollow center of our orb, it makes sense that when he came out through the polar cavities he could already see the sky and the stars, but making all those jumps, as he expresses them, only they are compared to a participation in an event that decades ago we would have considered "science fiction".

The prophet noted further: «*From there I was transported to the northern extremity of the earth and great works were shown me: I saw three doors of heaven open; through each of them come the spirit-winds from the north and when they blow there is cold, hail, frost, snow, dew and rain. If they leave through only one of the doors, they blow for good; but when they blow through the other two it is with violence and calamity on the earth because they blow with force. And from there I went to the western extremity of the earth and saw three gates of heaven open, the same number of gates and exits that I had seen in the east."* (1st Enoch 34 to 35) All this reinforces the idea that the atmosphere is being contained by invisible forces, powers and the simple elements. But in order not to go on too long with these cases of Close Encounters, let us put another part of his story: «*From there I was transported to the southern extremity of the earth and there its three open gates of the south wind were shown to me: for the dew, the rain and wind. And from there I was transported to the eastern limit of Heaven and I saw the three eastern gates open the three eastern gates of Heaven and above them small gates Through each of these small gates the stars of heaven pass and run along the course marked out for them to the west. Seeing this, I blessed the Lord of Glory all the time, and I will continue to bless the Lord of Glory, who has performed great and magnificent wonders to show the greatness of his work to angels, spirits, and humans, so that they can praise that work, all his creation, so that they can see the manifestation of his power and praise the great work of his hands and bless him forever.*» (1 Enoch 36:1-4). In all these cases he refers to heaven as "rakiyah", a Hebrew word that comes from "rek" (empty), not "shamaim" which is the classic definition to refer to Heaven, as the divine abode. Therefore, Enoch is identifying many different places, which can easily encompass the regions of our famous constellations.

That is why it is good to apply these ideas to references to Heaven or the Heavens: « *I saw a burning fire, and beyond those mountains is a region where the great earth ends, and there the Heavens culminate.*» (1 Enoch 18:9-10). If it says that "the Heavens culminate", it means that they have a geographical "limit" of some kind, outside of which there is still more. Enoch also speaks of the limits of Heaven and its foundations: " *Another angel who went with me spoke to me, revealing to me what was hidden, the beginning and the end, in the heavens above and in the depths under the earth, in the extremities of Heaven and in its foundations ...*" (1 Enoch 60:11) If there is any way to reach those heavens, will it be with spiritual elevation? With the Enlightenment? With a rocket? With a UFO? Or with a star portal? What is clear is that it is not about moving at thousands of km/h since we would not take a single step, not even at the speed of light, because it would take us centuries to go from one point to another in the universe. It is therefore obvious that trans-dimensional vehicles are used, both to generate Element 115 and 116 (Robert Lazar spoke about this after working in S-4, in Area 51, with UFO engineering) causing the curvature of space-time, to go from one dimension "immediately" to the other. Baruch wrote: «*And I asked the angel, my Lord, what is this sound? And the angel said to me: Even now the angels are opening the 365 doors of heaven ...*" (3rd Baruch 6:13) Is it not a coincidence that Heaven has 365 doors, the same as the days of the year? Is it not rather that these "doors" are the itineraries or locations in relation to the Sun? Since it obviously fits with the period in which our planet returns to the initial point after traveling around the Sun. Thus, the concrete of Heaven, one of them, assumes that it has the same magnitude as the Earth's trip in one year, so that full size would be from Heaven, which in that case is competent.

« *To the one who rides on the heavens of heavens, which have been from ancient times; Behold, he will give his voice, a mighty voice.*" (Psalm 68:33) As I have already said, it is important to take into consideration the fact that sometimes they speak of Heaven, other times of Heaven, and other times of "the heaven of heavens", as in the case of Enoch's journey, when Miguel shows him the sacred dwellings: "*He moved my spirit into the Heaven of Heavens and I saw that there was a crystal building and between those crystals, tongues of living fire."* (1st Enoch 71:5) Moreover, before the Deluge, Enoch had a dream in which he saw the high dwellings shake: «*In the year 500 on the 14th day, of the 7th month, in the life of Enoch, I saw that a great earthquake shook the Heaven of Heavens and the armies of Elion (the Most High), and the angels, thousands of thousands and myriads of myriads were shaken by the great agitation.*» (1 Enoch 60:1) It is understandable that the description of "Heaven of heavens" was used, exposing that it is greater and superior and that it is far beyond, even in essence and characteristics. That is why it is understood that there will be a new heaven, but not necessarily the elevated one, but the one that was placed on Earth so that those who would watch over it and help man would dwell there: «*"Then the First Heaven will disappear and pass away, and a new Heaven will appear. new Heaven and all the powers of Heaven will rise shining eternally 7 times more.*» (1st Enoch 91:16) There are those who consider that the description of the 7 and/or 10 Heavens of the Talmudic traditions is nothing more than 7 to 10 dimensions existing in this universe, while others see them as the orbits of our neighbors. Other postulates suggest that the existing dimensional planes are 11, and below the 3rd there is basically nothing, and celestial references would share a certain reality with this fact.

Well, what does the Scripture say?: «*Then Seth arose and came to his mother [Eve] and said to her: 'What is your problem? Why are you crying?' [And] she said to him: 'Look and see with your eyes the Seven Heavens open, and see how the soul of your father [Adam] is on his face and all the holy angels are praying on his behalf and saying: 'Forgive him, [oh] Father of all, for it is Your [living] image'.*" (Revelation of Moses 35:1-2) Why do some speak of 7 and others of 10? Is it possible that all 7 are in our system and the other 3 are much further away? The Apostle Paul also recounted having been taken, he or someone he knew, to other heavens: «*I know a man in Christ, who 14 years ago (if in the body, I don't know; if out of the body, I don't know; God knows it) was caught up to the Third Heaven. And I know of such a man (whether in the body or out of the body, I do not know; God knows), who was caught up to paradise, where he heard ineffable words that man is not allowed to express."* (2nd Corinthians 12:2-4) According to Paul, Paradise would be in the Third Heaven. And in another text of his, called the Apocalypse of Paul, he details other aspects of what exists Above, saying that he was raised, not only to the Third Heaven, but beyond the Seventh. It is possible that the conception of the Third Heaven is a cultural generalization to refer to the "Heaven of Heavens", where the Paradise of God is, since Enoch himself said that he had been raised beyond the Third Heaven, almost reaching the throne of Majesty, who is in the 10th Heaven. Although, in his description, Saul of Tarsus, the Apostle Paul, saw Above how a man was punished and then given another chance. This happened when Paul ascended to the Fourth Heaven. Then in the Fifth Heaven he said he saw angels making weapons for battle, and after the Sixth, he reached the Seventh Heaven and saw an old man sitting on a throne of light. In the end he observes the Ogdoad in the Eighth Heaven, and passes through the Ninth and catches a glimpse of the Tenth. Whether this text is true or false, the point is that trips to other heavens and other dimensions were very recurrent among the Israelites, as in cases where other people from other towns also experienced them, whether it was, as Paul said, in the body or outside. of him (in the Corpus Hermeticum it is given to understand that Hermes Trismegistus himself was led, at least on one occasion, to also contemplate the Ogdoad).

In the apocrypha we find that it is said: «... *We will enter [the first heaven] through it, and entered as if born from the wings, a distance of about 30 days of travel. And he showed me heaven in a plain, and there are men who dwell in it, with the faces of oxen, and the horns of deer, and the feet of goats, and the thighs of lambs. And I Baruch asked the angel, let me know, I pray you, what is the thickness of Heaven in which they traveled, or what is its measure, or what is the plain, so that I can also tell the sons of men? And the angel, whose name is Famael, said to me: This gate that you see is the Gate of Heaven, and it is as great as the distance from Earth to Heaven"* (3rd *Baruch* 2:2-5). So these vehicles exist and considerably they have been well used by the angels. Talking about journeys of 30 days and 180 days follows a pattern similar to the lunar cycle, basically the same as the terrestrial months – another curiosity. Famael told Baruc, in his case, that such a sky is as big as the distance from Earth to Heaven, which tells us that going through that sky completely would take 30 days, possibly in the vehicle they used, even being angels their crews. In the space or dimensional journeys of the prophet Enoch, he speaks of having been taken to visit and observe ten places that he defines as "heavens." planets? Universes? Dimensions? Regions? Heavens? Or orbits? He says that he was transported to the existing heavens. In the first he sees the place where the angels lead; in the second, where there are places of prison for the great deserters; in the third, places of eternal punishment; in the fourth, everything that includes the work of the sun, as if it were a machinery or gear dependent on all these ministers; then the fifth, where he sees the phoenixes (called "Bennu" by the Sumerians and Egyptians, "Feng" by the Chinese, and mentioned by the Greeks) but mainly observes the prisons of the great fallen, which are not compared to the chained deserters in the Second Heaven, those that he also observed; in the sixth, he saw the great angels and archangels, the ministers of heaven and protectors of Earth, the entire command organization of the children of heaven; in the seventh, he saw other angelic beings even more fierce; in the eighth and ninth he saw what is meant by the "path of the constellations." Would I then be having all kinds of star travel with these men?

Another example of the journeys of Israelite apostles and scribes is this: « *And when I had learned all these things from the archangel, who understands, he took me to a Fourth Heaven. And I saw a featureless plain, and in the middle a pool of water. And there are a multitude of birds of all kinds in it, but not like the ones [there are] here on earth. But I saw a crane as big as big oxen, and all the birds were bigger than the [biggest] in the world.* » (3rd Baruch 10:1-4) The Israelite prophets Baruch (Baruk or Baruch) and Zephaniah spoke of some of these places (although only Enoch cited the Tenth Heaven or Throne of God): « *And the angel took me there, to a Fifth Heaven. And the door closed. And I said, Lord, why isn't this door open so we can come in? And the angel told me: We cannot enter until Michael arrives, who has the keys to the Kingdom of Heaven, but you have to wait and see the glory of God. And there was a great sound, like thunder. And I said, Lord, what is this sound? And he told me, now Miguel arrives, the commander of the angels, he is reduced to receiving the prayers of men. And behold a voice, "that the doors will be opened".* » (3rd Baruc 11:1-5) These mentions lead us to think that the universe we know, the little we know of it, is really nothing based on what really exists. In fact, it was believed for a long time that most of space was empty, but, as I said before, it is now known that it is charged with electrical particles and plasma, that is, 99% of the universe is energy, and it even generates sound. and forms of life have been seen swarming in it -according to what I have heard-, forms that simply wander through outer space, without any ship, being like animals that roam in the immensity at almost Absolute Zero temperatures. Already in that order, species of flying snakes have been recorded on video in the stratosphere, with a very strange and unknown texture, somewhat whitish and apparently not very dense.

Thoth refers again to something interesting that reminds us of that Heaven placed above us: «*Seven are the mansions of the house of the Mighty One; Three guard the portal of each house from darkness; Fifteen paths leading to Duat. Twelve are the houses of the Lords of Illusion, finding four paths, each one of them different.*» (Table 14. Emerald Tablets of Thoth) These patterns are very significant. Although these characters had nothing to do with the history of Israel, they were aware of things: «*Look also at the hierarchy of the seven heavens, beautifully created in an eternal order and completing the centuries in different courses. Everything is full of light with no fire anywhere: for friendship and the combination of opposites and dissimilarities became light, and they shine on us by the energy of God, generator of all good and chief and conductor of the entire order of the seven heavens.*" (Intelligence to Hermes. Treatise XI, verse 5-7. Corpus Hermeticum) In Taoism there is talk of the existence of 9 kingdoms and 3 heavens (also in Nordic culture it was believed that there were 9 kingdoms). A striking aspect in Chinese culture is that they call the universe "Tian-Di", which literally means "Heaven and Earth". Although, although in Taoism there is also talk of several heavens, in a certain trend in China they believe specifically in 9 (although they could be more specifically the 9 kingdoms of Taoism, as in Nordic culture), which would be cared for by each of the 9 daughters of Yu-Huang and Wang Mu Niang-Niang. As far as Taoism is concerned, I would affirm that these heavens are only 3 (called San-Qing, which means: "three pure"; at the same time the name of 3 deities that inhabit them) and they come to be governed by venerable emperors (Tian- Zong). The first would be Yu-Qing (Heaven of Jade Purity), inhabited in turn by Yuan-Shi, part of these 3 Tian-Zong. That sky would also be governed, according to traditions, by the Jade Emperor. The Second Heaven would be called Shang-Qing (Heaven of Great Purity), reserved by Ling-Ba or Ling-Bao. The third, called Tai-Qing (Heaven of Highest Purity), is ruled by Tao-De – another reference defines that place as Da-Luo-Tian, where Yu-Huang, the Jade Emperor, dwells. According to the predominant religion, Si-Ming, the Lord of Destiny, would be a reviewer and recorder of the deeds of men, noting their actions in the Book of Death and the Book of Life –a striking peculiarity.

Although the heavens were defined in 3, as in the Bible, the parallels go further, since they consider, like the mention in the book of Zechariah, that 4 great spirits (Ssu Ling) administer the 4 angles of the Earth. They also talk a lot about spirits and, specifically, about the fight against demons since time immemorial. In a similar way it can be said of the version they have of Creation, where clearly divine beings intervened in the separation of Heaven and Earth, as in the case of the fire god Li (mainly called Zhu Rong). They believe that in the origins everything started from a primordial chaos and dark waters, as well as in the Flood, and the Abyss that is located under the crust of the world. In the same way they follow the thought that in the origins there was a great monster, a demon called Gong Gong, against which the deity fought and defeated. They also talk about the "foundations" of the Earth, the supports of the cardinal points, basically that the Earth and Heaven are supported by pillars, as the prophet Enoch refers to. As I mentioned before, they teach that Pan-Gu was born as a cosmic egg, forming Heaven (Yang) from the upper part and Earth (Yin) from the lower part, according to the myth, growing as: 10 feet per day (the Legend assumes that he divided Heaven and Earth with his hammer, giving great importance to this weapon, like Thor in Scandinavian culture). It is said that he extended his shell more and more after 13,000 or 18,000 years, after which he himself was seen. His eyes would be the Sun and the Moon, his head the 4 great mountains, his blood the seas and rivers, his hair the grass and the trees, his breath the wind, his sweat the rain and his voice the thunder. In this way, the fleas on his body would become the ancestors of humanity. I think all of these elements are remarkable in their similarity to the enormity of examples and symbols from the many other generically referred to cultures.

Although, for the Egyptians there were also broad definitions of heaven. They regularly described the above as divided into 3 parts: 12 celestial bodies in each of them. These 12 regions were equated with 12 dogs, possibly guardians, which had 12 dogs in turn (adding 144) – hence the day was also divided into 12 hours (12 + 12). Thus it is very similar to the distribution of power and the creations made by Sakla, trying to duplicate the Kingdom of Christ in the imperishable places. These may be recognized as "dogs" (Revelation 22:15), and are also the ones mentioned in mythologies, as guardians, either of the dimensions or of Hades, as reflected in the Greek legend of Cerberus or Cerberus. Certainly, dogs have always been a symbol of the protection of important places, as much as dragons or fabulous beings. But in the realm of the "beyond" they seem to be vigilant posthumous dogs. This could be the reason why Anubis, the son of Nephthys (Nebt-Het), is custodian of the underworld and is described as a man with the head of a dog (specifically it could be a type of jackal or a greyhound or black hound), and this too would explain why the sons of the Norse god Loki were given a terrifying description as the destroyers of the world who have been driven from the inhabited realms. Consequently, the idea of watchdogs is widespread and popular.

In important locations, dragons were formerly placed to watch over them, to protect relics, but the secondary mammals were these paths. The clear example is how they are fixed in Heaven to protect it and scare away demons, according to Taoism (in this Asian belief it is the guardian Er-Lang who directs them). These celestial hunting dogs are known in Chinese culture as the Tian-Gou, although these celestial dogs are also often referred to as negative, in the case of the persecutors of the male gift Zhang Xian, where, under the same name, they are defined in the singular as "greyhound of the skies". Note that in Hebrew, dog is "Keleb" ("Je", "Ke" or "Kel" = "likeness"; and "Leb" = "source of..."). It is something similar to: "like the conscience", "like the inside", "like the center", "like the heart", while "Kel" is associated with "captivity" or "enclosure". This may be why they were animals to guard or enclose mysteries, since the very word "Leviathan" also has a similar etymological root: "Leb-i", "ha-tán" or "titén", that is to say: "the heart of the dragon", "the center where those are". However, these dogs do not seem to guard the heavens as such, but simply the connections between dimensions or realities. Where did they come from? Who put them there? We have to refer all this to the later points, since we have to understand that everything that exists in these realities arose at the time in which Creation took place from the imperishable powers, coming from his Kingdom. This spiritual place brought life forms to a mental universe, where everything that governs the forces of our world and other existing ones was expressed, such as the "laws" of nature, physics, quantum mechanics, etc. So, life on the physical plane has not progressed by chance or chance, but by the action of entities interconnected with created things, which could be the reason for much of Eastern thought, especially animism, which considers that all are invisible forces and ancestral spirits.

Ergo, let's clarify that astral travel is not recommended, and in that order, the prophets did not provoke them, but were deliberately invited by God's messengers. I affirm this because precisely leaving the body to travel without it is taking away the body weight to move anywhere, but it is dangerous, because entities abound in those planes that hunt down every form of life that passes by: «*Listen, oh man, to the depth of my wisdom. I speak of the hidden knowledge of man. Far have I been in my journey through SPACE-TIME, even to the end of the space of this cycle. Yes, I caught a glimpse of the HUNTING DOGS of the Barrier, lying down waiting for whoever passed by. In that space where time does not exist, I faintly felt the guardians of the cycles. They only move through the angles. They are not free of curved dimensions. Strange and terrible are the HUNTING DOGS of the Barrier. They follow consciousness to the limits of space. They do not intend to escape by entering your body, since they quickly follow the Soul through the angles. [...] Once, in the past time, I reached the great Barrier, and I saw on the shores where time does not exist, the shapeless shapes of the HUNTING DOGS of the barrier. Yes, hidden in the middle beyond time I found them; and THEY, smelling me in the distance, rose up and their howl could be heard from cycle to cycle and move through space towards my soul. I then quickly fled before them, back to the unthinkable end of time. But they always followed me, moving at strange angles unknown to man. Yes, on the gray shores of the end of SPACE-TIME I found the HUNTING DOGS of the Barrier, ravens for the Soul that tries to go beyond.*» (Emerald Tablets of Thoth. Table 8)

The Foundations

«Yours are the heavens, yours also the earth; The world and its fullness, you founded it." (Psalm 89:11) Now the reference to the "foundation" of this dwelling is becoming clearer. As with Moses, there are reports about what was done in that interval that later received the name of Yom Sheni (translated as Second Day). Aaron's brother said that the second period was identified by the appearance of the heavens, after the waters had parted. In this chapter God begins by saying "kabaati", which is similar to "kba" (to defraud) or "kebaat" (cup or chalice). The translator put: "firm". We can deduce that, since the glass has a semicircular shape, and some of them are diaphanous, it is used as an example to elucidate what we see above us as Heaven, assuming that it was the right

word to refer to the appearance of the atmosphere. He then affirms that the waters were "aligned", taking the Hebrew "cav" which is "rope", sometimes used as a synonym for "line". This could imply that a uniform and smooth appearance was given to the water, as if a string were stretched. Those that were under the Shamayim had to be put in a single group and the "heaps" dried. The Hebrew word "galeihem" speaks of lots "of them", referring to bodies of water ("hem" is "they"). The word "gal" is used in Genesis 31:46 as "gal avanim", consonant with "heap of stones". Similarly, in Isaiah 48:18 it seems to speak of "wave", of the "waves of the sea". A little later he emphasizes that the rock is "laqídeti" dryness. The voice "lacad" can mean more than one thing, for example, "to catch", "to capture", "to take by lot" (Amos 3:4, Judges 7:25 and Joshua 7:14). It also refers to "being caught", "being captured", "being taken by lot" (Isaiah 8:15). In another case he speaks of "hardening", "cohering" (Job 38:30). It also speaks of "clogging up" (Job 41:9/17) or identifies a "trap" (Proverbs 3:26). Likewise, Enoch uses the word "asafti", regarding the sea. Asaph, like the name of King David's singer and scribe, means multiple issues: Collect, reap (Deuteronomy 16:13 and 28:38). It also refers to storing food (Genesis 6:21). Identify collecting or collecting money (2 Kings 22:4). It can refer to gathering people together (Genesis 29:22). It is used to "collect", "bring someone home" (2 Samuel 11:27 and Psalm 27:10). You can identify withdrawing or taking away (Jeremiah 16:5 and Psalm 104:29). It is used to say "turn away", "withdraw your hand", that is, "stop doing something" (1 Samuel 14:19). There are cases in which it is "to assemble" (Deuteronomy 33:5), "provisions", as in the case of "bet ha-asum" = "house of provisions" (1 Chronicles 26:15). Or it is also "harvest" oriented (Isaiah 32:10).

For its part, "sof" is "final", which can be its "term": « *He founded the earth on its foundations; It will never be removed. With the abyss, as with a dress, you covered her; Over the Monts it was the waters. At your rebuke they fled; At the sound of your thunder they rushed; The mountains climbed, the valleys descended, to the place that you founded them. You have set an end for them, which they will not cross, nor will they cover the earth again."* (Psalm 104:5-9) When it says that he "bound" the seas by "sand", it is taken from the Hebrew "ekshrehú", whose root "kesher" (like the sky god of Sumerian mythology, Kishar) refers to bind, bind, unite or conspire. For its part, "jol" can translate: "go around" or "in succession", "irrupir" or "dance" (go in circles), although more specifically it can be translated "sand" or "ave Felix". God says that in the middle of the Earth he made a hole, an abyss, and also limited the seas "giving them their perpetual statute". In the same way, he pointed out this as the beginning of his Creation, made by Him and for Him. Therefore, that reminds us of the words of the Apostle Peter in this regard: « *These are voluntarily ignorant, that in ancient times they were made by the word of God the heavens, and also the earth, which comes from water and by water subsists, for which reason the world of then perished drowned in water ...*" (2nd Peter 3:5-6) It seems that each period is accompanied by a new "tomorrow", as two separate things: « *And God said: Let the waters that are under the heavens be gathered in one place, and let the dryness be revealed. And it was like that.*" (Genesis 1:9) If the water that remained under that atmospheric sky came to settle below, gathering, it is because it was not unified before. It is possible that the mass of water was not condensed or stationed in a single region, in a compact way, and that enormous water, already divided, went up and down, the bottom going to concentrate in the same place. If this idea is true, we would have the primordial Earth as a place, Aretz, which has a dark hole (Tehom) and waters that are divided, putting some above the apparent atmosphere and others below, but where exactly below? Let us keep in mind that many biblical passages warn that there are waters on the surface of the Earth and waters under the Earth, since everything is measured and adjusted in detail: "You *set all the ends of the earth ...*" (Psalm 74:17) Enoch refers to this in the following way: « *All the sea, all its waters and all its movements, are they not the work of the Most High, has He not put his seal on all its [sea] action and has not completely chained it? to the sand?* » (1 Enoch 101:6)

So there comes another overwhelming aspect for those who fully immerse themselves in the vision: « *discover the dryness* ». What is dry? Is the text implying that planet Earth was a huge body of water with no glimpses of sand? Now, was there dry land under all that amount of liquid? It is telling us that there was a change of colossal magnitudes in the terrestrial geography, resulting in a considerable geological modification, similar to that exposed in the theory of Pangea (the primeval super-continent, which is said to have divided, advancing slowly until showing the map that we see today in satellite photography). That is why Pangea means "all the earth" (pan-gea, in Greek). Again, continuing, he prays: " *And God called the dry land Earth, and the gathering together of the waters he called Seas. And God saw that it was good.*" (Genesis 1:10) The scholars wrote "earth" in this part, specifying that it is "terrain", not sounding like Aretz but rather Eretz (land, territory, land in association with dry and firm aspects of a region), although the letters are the same three: Alef, Reish and Tzadik (whose composition represents the action of God leading an event to manifest his justice). Grammatical "dots" were not added to Scripture until fairly recently (it was originally read without them, since Hebrew does not have vowels). So, the dry part received the name of Aretz or Eretz? The Sumerian tablets, which are the oldest officially recorded records in existence, say that the Earth was almost entirely covered by water, swampy, humid parts, and deep lagoons, until arid, dry regions of firm ground and hard rock began to stand out, where it was even possible to build a civilization. That site, as I said before, received the name of Eridu, from where the Hebrew Aretz and Eretz come from. If this is related to what we are referring to, the name Aretz would have been the global definition of our world, although it was once only the dry region, which makes sense, since our planet, although it is about 70% water, is called ironically "Earth".

So at the beginning, when he says that he made the heavens and "the Earth", he makes it clear that he is not referring to its situation but rather defines it as the place (the name with which it settles and is already known), regardless of whether it was watery, chaotic, formless, abysmal, or dark. It is in this order that the distinction between what is firm land (Eretz/Aretz) and the rest of the liquid mass comes, an aqueous quantity which is wisely defined as "Yamím" (seas. From the singular "Yam") since these two Hebrew initials "Y/I" and "M" are the masculine plural in this language. In other words, they are mass, so it can also be understood as "crowds", "people", etc., although many times the Yod (I or Y) is not observable or pronounceable in the writing of a word, as in the case of "Abraham" (father who will see multitudes). Although, in a cuneiform carving, it speaks of the great primordial gods, which have never come down to Earth. Why have they never come down? These same complete 7 important divinities, next to the sun god and moon god or goddess. Once again we find that ancient myths recounted things in the way their memory had remained, as representative symbols. The 7 great gods would be Mars, Mercury, Venus, Jupiter, Saturn, Uranus, and Neptune, which can be identified by their other European names: Ares, Hermes, Aphrodite, Zeus/Amun, Cronos, Uranus/Heaven, and Poseidon (to Pluto), or Hades, he was not considered in principle, although he has his own interesting history). The sumerologist Zecharia Sitchin also had this hypothesis that I have just referred to, being a scholar in the study of Mesopotamian writing and culture. Sitchin commented in his book "Genesis Revisited" that the waters of Tiamat, from which the Earth arose, separated from it, creating another planet. Something similar would have happened with other bodies in our system, thanks to the collapse of certain orbs that had not yet defined their orbit. According to Mr. Sitchin, Tiamat would be the abyss that Moses referred to, the Tehom, a quite logical aspect. The name TMT –excluding the vowels- is similar to THM, to the extent that the plural of Tehom, that is, the abysses, would be "Tehomot (THMT), as Exodus 15:5 points out. A small change in aspects of linguistics after the process of time and displacement would give a briefly different verb form but with the same root (Sumerian Tiamat is the same as Hebrew Tehom). Thus we could find these "waters" above the atmospheric layer in other bodies of our solar system.

The version of the scribe Ezra, in this regard, is this: « *"The third day he commanded the waters to gather in a seventh part of the earth; six parts to be dried up and preserved so that some of them could be planted and cultivated and of service that you have before you. By his word they went out, and at the same time the work was done. Immediately came the fruits in abundance and endless varied appeals to taste, and flowers of inimitable color, ineffable and scents of perfumes These were made on the Third Day.»* (4th Ezra 6:42-44) Ezra was notified that the waters that did not rise were to settle on 1/7th of the Earth - although technically they occupy much more than that, possibly because their quantity is spoken of on the surface, not inside the globe. Once this action was carried out, the vegetable life forms could swarm, since without water, how could they do it? If from 7 parts one was chosen for the sea, and then, apart from the Earth, 7 liquid globes were created, this number is relevant Also, in reference to that "Third Day" or "Third Age", we find: « *And on the third day he commanded the waters to pass from the face of the whole earth in one place, and the dry land to appear.*

And the waters did as he commanded them, and he withdrew from the face of the earth in a place outside this Rakiah, and dry land appeared. And on that day he created for them all the seas according to their different meeting places, and all the rivers, and the gatherings of the waters in the mountains and on all the earth, and all the lakes, and all the dew of the earth, and the seed that is sown, germination and all the things, and fruit trees, and trees of wood, and the garden of Eden, in Eden and all. These four great works that God created on the third day. » (Jubilees 2:5-8) Despite the fact that at first the sea was left in one area, in relation to the other "seas" that went up, now it was relocated limiting it in various sections and in various ways , making it clear that, if only one of the 7 parts is what we see, there must be aquifer deposits of a supernatural size throughout the interior of the Earth – this would explain chapter 6 of Genesis, when it says that the water stores are they let it release, flooding the planet in a few weeks (rain wasn't really to blame for the world's flooding).

Once knowing how the atmosphere was formed, let us add the data that assumes that our world is subjected thanks to a very specific point that shows to be located in some specific location below the surface, which is defined as "Cornerstone", without which the world would not be supported: "*and I also saw the cornerstone of the earth and the four winds that support the earth and the firmament...*" (1 Enoch 18:2) All the accumulation of literary works of yesterday do nothing but leave us understand the majesty and power of God: " *O Lord Jehovah! behold, you made heaven and earth with your great power, and with your outstretched arm, nor is there anything difficult for you ...*" (Jeremiah 32:17) As a whole, this narrative, as well as the subsequent one, have also its explanation in the symbolic field, which is visible to a large extent thanks to the important works of Rabbi Félix Guttmann Van Katz. Although, the consequent point can be supported by the texts of Nag

Hammadi, which spoke about the seed of Eros (spirit of passion) absorbed by the Earth. This would give rise to the following aspect of the story: « *Then God said: "Let the earth produce green grass, grass that bears seed; tree that bears fruit after its kind, whose seed is in it, on the earth." And it was so.* » (Genesis 1:11. RVA 60) By always saying that "God said" it must be understood as "the deity" (Elohim), not necessarily personifying it under any label. It is understood that there are deities subordinate to the Father in our reality, which obey the Holy Spirit, Christ and the eternal will of the Father, with which a specific name is not defined in any part of this matter, until the appearance of Jehovah in Genesis 2:4, which is attributed to the personification and designs of God: «*Lord, you are worthy to receive glory and honor and power; because you created all things, and by your will they exist and were created.*" (Revelation 4:11)

So the earth began to produce plants, apparently, without any human intervention, the creative power being patent in the landmass itself, which gave all kinds of results, both in fruit trees and plants, flowers and shrubs, having its own codification. already established, which indicated that, when the surface bears fruit, it had the intrinsic properties of producing again, with complete information already grafted, which we call "seed". The Barashit marks this stage as the production of the "green" (desheb esheb), translated as "green grass." The literal text says, more or less, something like this: «*And Elohim said: let the Aretz produce green grass from its seeds, the seed of a fruit tree that bears fruit according to its type (species), whose seed is in the Aretz, and it was like this.*" So two things arise: simple plants and fruit trees. Something like vegetation without more and also products that assumed the role of bearing fruit in themselves (one is grass food and the other is fruit and vegetable food). Obviously, the stage is set here for those who eat both: grass or grass and fruit. That is to say, even if the destiny or design of the aforementioned things were concrete, this already points to the consideration of a self-sustaining ecosystem, suitable for the life of developed creatures, as forms of intelligent life -with which I obviously include, to the animals. Ergo, as the following verse (12 and 13) points out, everything said by Elohim happened exactly like this, resulting in the pattern of Yom Shlishí (third day). Said aside about the plants, it did not stop being mentioned by Enoch: «*And so I made all the heavens; and it was the Third Day; And on the Third Day I ordered so that the Adamah would abound with large and prosperous trees and the mountains, all of sweet grass and all seed (seed) on what was scattered; and I put a garden and closed it, and I guarded [with] malají of fire nurturers-attentive. And so I did honor to the world-universe. And it was late and it was morning Fourth Day.*» (2nd Enoch 30:1-2) This version differs only in that it adds the fact that by then its beautiful Garden and its custody had already been established.

The Planets

You have to see a matter about plants: they need air, sunlight, soil and water. At no time is the appearance of oxygen for the respiration of life forms mentioned, but it is assumed that the atmosphere would have this characteristic. It can be theorized that when the said "waters" were divided and the said "firmament" appeared among them, including the apparent condensation of said liquid masses, the atmosphere was full of oxygen molecules. It is evident that if the component particles of water are separated (2 molecules of hydrogen and 1 of oxygen) we would have breathable air. If this is correct, we would understand that life was not viable until the air was subtracted from the water, giving another location (planet of this system, for example) for the extra hydrogen released. It is also possible to "hypothesize" that, since we would not know the nature of the extra hydrogen -suppressed to give rise to oxygen-, it could be carried along with the other waters, possessing a greater amount of H molecules. Even accepting this possibility, These masses, now with a greater amount of hydrogen, could have been subtracted from the necessary amount of oxygen for our world or for others, such as Mars or Venus. On the other hand, to harbor plant life there would still be a lack of sunlight. With everything and that 3 ages have passed, the Sun is still not mentioned, how then could green grow? In the Book of Jubilee the following is related: «*And on the fourth day he created the sun and the moon and the stars, and set them in the expanse of the heavens, to give light to the whole earth, and day and night, and divide the light from the darkness. And God appointed the sun to be a good sign on earth for days and for Sabbaths and for months and for celebrations and for years and years of Sabbaths and for anniversaries and for all seasons of the year. And that separates the light from the darkness [and] from the prosperity, that all things that can prosper and grow shoot upon the earth. These three guys made the fourth day.*» (Jubilees 2:8-11) The fourth stage becomes the appearance of the lights. They follow organized patterns, fixed sequences, concrete precessions and thus determine the cycles for all times. Since we are not talking about the galaxy or the rest of the universe, it is appropriate to consider that it is talking about this part of the Milky Way, technically up to where it covers the maximum distance of the constellations that we see. That means that they were located in the "expansion of the heavens", since they are not in our atmosphere, but visible bodies at "galactic" distances.

Enoch also mentions the account of the defined Third Yom, in his case thus: « *And for all the armies of the heavens I imagined the image and essence of fire, and my eye looked towards the very hard, firm rock, and from the flash of my eye the lightning received its marvelous nature, which are both fire in water and water in fire, and one does not put out the other, nor does the one dry the other, therefore the lightning is brighter than the sun, softer than water and firmer than hard rock. And from the rock and I cut a great fire, and from the fire and I created the orders of the armies of tens of thousands of uncarnal malachim, and their weapons are fire and their armor a burning flame, and I commanded that each one should stand in your order. And one of the chief custodians of the malachim, stubborn (twisted) with the custody downwards and promoted an impossible plan: and erect his throne above the Earth considering his strength compared to mine. And the consequences of his puffiness with his malachim and he was prosperous unfortunately on the face of the Abyss, always.»* (2nd Enoch 29:1-4. Conventional translation) Contrary to the religious idea about angels as gaseous beings, God taught Enoch that the principles of their nature come from fire (if we look objectively, humans are mostly water, but this is not perceived with the naked eye, so another element, such as fire, would not have to represent a burning being). The reason for mentioning the beginning of the existence of lightning reminds us of the figures that ancient peoples conceived as most important, since the possessor of said "weapon" was understood as the supreme god. Making use of this triple power (because it incorporates lightning, thunder and lightning), throwing it against solid stones is how he caused that "fire" from which he molded his countless hosts. The last part, referring to the "obstinate" cherub, will be worked on in the next work.

It is notorious that he considered these beings "lo bakar" in Hebrew, that is, they do not belong to the prototype of human flesh, or simply to its very substance. Another aspect is the descriptive way of referring to said entities, since the common way of uniting the idea of "fire" and "angel" is based on the definition of "seraf". Seraphim, as they are known in Spanish, are not tiny celestial messengers with feathered wings and the likeness of naked babies. Saraf literally means "fiery", and is a synonym for "snake", as Isaiah points out (chapter 14:29), when speaking of the *"flying serpent*", which in Semitic is written as "saraf meofef". The case is lucidly appreciated in the account of the "saraf" (translated as "fiery serpent" in Numbers 21:8) that Jehovah commands Moses to "manufacture" and which is made a "nachash najeshet" (translated as "serpent of fire"). bronze" in verse 9, and a similar example found in 2 Kings 18:4). It is not surprising the enormous similarities that link the angels with the stars: «*Observe all the things that happen in the sky, how the luminaries of the sky do not change their path in the positions of their lights and how they are all born and set, ordered each according to its season and do not disobey its order.*» (1 Enoch 2:1) It is evident that the stars follow their courses, but they do not make decisions by themselves. They all obey patterns and are guided by celestial messengers who are identified with them: «*Lift up your eyes and see who has created these things. He brings out and counts the army of them; He calls everyone by name. Due to the greatness of his vigor and the power of his strength, none will be lacking.*" (Isaiah 40:26) While Isaiah refers this about them, David wrote: "*By the word of the Lord the heavens were made, and <u>all the host of them </u> by the breath of his mouth. He gathers together the waters of the sea as a heap; He puts the abysses in deposits.*" (Psalm 33:6-7)

Ezra cites the case thus: «"*On the Fourth Day in which he commanded the brightness of the sun, the light of the moon, and the disposition of the stars to come into force; and he commanded him to serve humanity, about to be formed.*» (4th Ezra 6:45-46) The influence on us of the observable stars (the closest ones) is perennial –it is not a mystical belief of esotericism, but a proven fact. Astrology is not acquisitive to the readers of the Tarot, since it was an element taught by angels to humanity, and with time completely distorted. I do not know the procedure and work of the stars that are beyond the limit of what the framework of the location of our constellations encompasses, but these, those that concern us, have a job of connecting with us, not only as stars but being guided and represented by celestial messengers. That is why the Moon, not being a star, reflects sunlight, and in the spiritual sense it is also a light: "*And Elohim said, Let there be lights in the firmament of the heavens to separate the day from the night; and let them serve as signs for the seasons, for days and years, and let them be lights in the firmament of the heavens to give light upon the land. And it was like that.*" (Genesis 1:14-15) Then "lights" appear, of what kind? The first thing that comes to mind are "stars", with which we would be assuming that the Earth is much older than, at least, the suns that we can observe. It is noteworthy that it says that these "meorot" (as it appears in Hebrew, and which comes from the same root, "Moré", which is "teacher") appear "ba-rakiyá ha-shamaim", that is, "in the emptiness" that had received the name of "heavens", which implies that its job, before man was there, was to look through him – in addition to diminishing the power of darkness.

Keeping in mind that the Hebrew plural voice "meorot" comes from the MR source, that is, "mere" or "meriá" (root of the name Miriam or Maria), whose meaning is "rebellion", as if on our planet it was imperative that Overflying forces of light in charge of guiding, protecting and bringing truth in the midst of the existing rebellion, or being part of the mix of events that we know as "The Rebellion". It may be simply a theory, however another fact supports my approach, and that is that the function of these "lights" is to " *differentiate between day and night.*" Although, how would the stars separate day from night? Day and night were already separated and differentiated in advance, remember? And it has been taught that what divides the span of both is the Sun itself, not the stars (the star king potentially influences everything, but the stars are simply observed as small dots attached to the text of the celestial vault). Certainly "stars" (Hebrew "kochab") is the regular designation to refer to God's messengers, who, on a spiritual level, separate the light from the darkness. In this regard Enoch also wrote: « *Because the signs, times, years and days were shown to me by Uriel, the angel to whom the Lord of glory has entrusted the direction of the lights, of Heaven and in the world, so that the sun, moon and stars and all the creatures that serve, that revolve on all the chariots of Heaven, reign on the face of Heaven and be seen on earth and be the guides of the day and of the night.*" (1 Enoch 75:3) What are the "chariots of heaven"? That was the way in which ancient civilizations had to call what we currently define as "UFOs". That is why it would not be strange to read Enoch's points like this: «*I have seen in the Heavens chariots that travel the world above those doors and in them roll the stars that do not hide. There is one greater than all, who turns the whole world around.*» (1 Enoch 75:8-9)

That special chariot or chariot is said to pull the Sun. The Greeks said that Apollo was the driver of the "chariot of the sun", once Zeus had burned the son of Helios (Sun) for passing very close to the Earth in the car solar energy and cause terrible global warming. In addition, although Uriel is not mentioned in the Bible, he is one of the most important archangels, being the head of the celestial angels in relation to the work of each one as luminaries, and also, as we saw before, he was the one whom God summoned to give rise to the divine world of light. Enoch wrote: «*From there I was transported to another place in the west, in the extremities of the earth; I was shown a fire that ran without rest and without interrupting its course neither by day nor by night, remaining constant, meanwhile. I asked saying: "What is this that has no rest?" Rauel answered me: "The function of this fire that runs towards the west is to guide all the luminaries of Heaven.*" (1st Enoch 23:1-4) According to this description, the Sun is captain of the stars of the bodies of our system, to which all must obey, that is, they are subject to the established laws that we know as orbital movements. For example, a fragment taken from the oldest Sumerian tablet says: "*The reptiles truly descended. The earth is resplendent as a well-watered garden. At that time Enki and Eridu did not appear. Daylight did not shine, moonlight did not emerge.*» I wonder if it was glowing and neither the Sun nor the Moon were yet, how come the surface glowed? This cuneiform inscription warns that at the time the reptiles descended to Earth the mythical gods had not yet arrived or left. had laid foundation of any civilization, yet with all the planet was beautiful and arranged, without even having Greater and Lesser Lumina. Some glimmer of light must exist then, at least from the high Heaven could glimpse even a dreamlike appearance in the crust, No?

In any case, there are still references in these chapters of Genesis 1:14-15, where it states that the objective of these luminaries was also to be "signs" (in Hebrew it appears: "le-otot", that is, to be a sign). The sign is a reference to designate something, remember it or expose a larger idea, as is a symbol, which is simple but in itself expresses a lot. The other peculiarity of these lights seemed to have something to do with the fact that they were "dwellings", since in the Hebrew language it says "la-moadim", since "moed" is an agreed meeting place (Joshua 8:14), it also identifies a fixed time or term (Exodus 9:5 and Daniel 12:7). This word also refers to a time of celebration and festivity (Lamentations 2:7 and Zechariah 8:19), it also refers to an agreement (Judges 20:38) or an assembly (Isaiah 14:13). For example, "ohel moed" means "tent of meeting" or "tabernacle" (1 Kings 8:24), and the phrase "bet moed" means "destined house", in this case alluding to Sheol (the abode of the dead, as Job 30:23 refers). It cannot be denied that this does not fit much with the idea of the stars but rather the place of assembly and military meeting of the angels and forces of God, designated to take care of human affairs. Moreover, "moad" defines people summoned for the military campaign (the RVA version puts "ranks"; thus, it translates "éin bodéd be-moadáv" as "there is no one left behind in their ranks", in consonance *with* Isaiah 2:31 p.m.).

Then the biblical passage from Genesis adds that another function of these "lights" is " *ve-le-iomim ve-shanim*." Which means that they would be there "for the days and the years", or "for the ages and years". With which this would serve so that they were of light on Earth while being in the Rakiyah of the heavens. Natural stars obviously do not seem to do that, besides the fact that the Earth is in the middle of them and not them on the Earth – applying the known laws of astronomy. The suns that we observe serve as a reference, signal or sign according to their positions (zodiac) and clearly this is a story in itself, but they do not separate day from night, since they are not observable during the day and only make an appearance at night (then what they would do would be to lessen the darkness, at most, unless in their corresponding area they did their own work, as our sun does with us). Likewise the stars can serve as chronological guides, according to the precession of the equinoxes, but they are not really light on Earth, unless they are referred to as beings of light who take care of human spiritual affairs in this world. Even another aspect that contributes to my approach is the following chapter: « *And God made the two great lights; the greater light to rule the day, and the lesser light to rule the night; He also made the stars*. (Genesis 1:16) With all this process, only now are the stars manifesting! So what luminaries were before? Now it is plausible that the deity created the Sun, the Moon and the rest of the luminous bodies of the cosmos, still breaking with the parameters established by official science that the earth is about 4,300 million years old (maximum 4,700 million), while the The Moon is up to 800,000 years older, and the Sun is supposed to be even older. In any case, this is not conclusive because what is being studied is the crust, the strata and other aspects, evidence that can vary with abrupt changes in the geological layers (hence the incredibly older strata found on the Atlantic ocean floor), for example, and then in the Pacific, than in the rest of the Earth's surface).

It is strange that God ordered to appear, first lights, then the Sun, then the moon, and, finally, the stars. They talk about 4 genres, with the exception of the Moon, how do the others differ? The Sun is a star, and the luminaries, if they were not also this type of stars, what would they be? So the sunlight appears in this 4th period, giving reason to the fact that the light that appeared at the beginning was not solar, nor are the days mentioned 12 or 24 hours long – dependent on the Sun. And this is all confirmed by his own words. of the proto-Hebrew prophet: «*First there appeared the great luminary whose name is Shemesh (Sun) and whose circumference is like the circumference of Heaven and is totally filled with a fire that illuminates and burns.*» (1st Enoch 72:4) If it were a form of reading that determines the order that there are light givers, which is expressed in the indicated stars, then there would be no confusion. Regarding the detail given by Enoch, we again observe that the diameter of Heaven corresponds to the "circumference of the Sun". Adding data we go to the Egyptian manuscripts of the Nag Hammadi Library, where it is stated that « *Just as the phoenix bird (phoinix) testifies about angels, this is also true in the case of water vessels that are in Egypt: about what they make a witness that will happen at the baptism of the true man. The two bulls are in Egypt, a hidden meaning: the sun and the moon. This is a testimony that they make about Sabaot, namely, that the Wisdom of the world has preceded over them since the days when the sun and moon were created and in which it has sealed their heaven for eternity.*" (Coptic Apocryphon of creation 122:16-26) Skipping the observations that these exposed verses cause concern, the initial part that relates the phoenix to the angels, reveals the parallelism in relation to fire.

David wrote: "*When I see your heavens, the work of your fingers, the moon and the stars that you formed ...*" (Psalm 8:3) It is evident that this narrative expresses great things in few words, using symbolic and literal references such as self regularly used when referring to a word and preconceived idea in society, which exposes something much deeper. The Sun and the Moon are also defined as "meorot" and their function is similar to the previous ones, but superior in all aspects. With which, they have a spiritual work, ruling over day and night, and also the practical part, which is natural light. If it says that our star king dominates the day, it is clear that it exists independently of said luminous body, because otherwise it would say that the Sun dominates the darkness. The day already existed -since we are going through the narration of the 4th itself-, but a great spatial light is placed to govern all the functionality of this system that prevails in the spiritual light, also being the clarifier of the surface for about 12 hours daily –according to Ecuador. The other thing is that the Moon dominates the night, absurd art, since Selene does not generate its own luminescence but rather reflects sunlight. In addition, the bouncing glow is contingent on its monthly displacement work, sometimes being almost invisible and other times totally visible in the dark. The night would need a king over it that would be more illuminating, like the Sun, and the Moon, already having luminaries at its side, would work in the already existing light, but it is not like that. So our natural satellite does little to diminish the darkness, while the light is more than powerful in itself. All this is symbolic, all this is spiritual, and must also be seen literally in due order. The luminaries that appeared follow celestial lines and correspond to stars in a certain sense. The two most important luminaries, as Enoch shows, are directed as enormous mechanisms that work thanks to the titanic work of a multitude of angels, who, in turn, are under the orders of various commanders, who in their own order respond to a kind of of "overall." For this reason the genesis account of these "days" concludes with: "*Therefore the heavens and the earth were finished, and all the host of them.*" (Genesis 2:1) Army?

The Sun and the Moon would then come to be "bodies", as we already called them, with members that maintain them, just like the fire brigade, the police corps or the Body of Christ. This is how we see the "kojabím" (stars) also appear, which can also be both: spiritual lights and natural lights. The generic work of these resplendent organisms was slowly pointed out by Enoch: «*Then I saw other lightning bolts and stars from Heaven and I saw how He called them by their names and they paid attention to Him. And I saw how they were weighed in fair scales, according to their luminosity, their dimensions and the day of their appearance and how their movement generates lightning; and I saw their course according to the number of the angels, and how faithful they are among themselves. I asked the angel who was with me and he showed me what was hidden: "What is that?" He told me: "The Lord of Spirits has shown you his parable; these are the names of the saints who live on earth and believe in the Lord of Spirits forever and ever.*" (1 Enoch 43:1-4) Although it was a parable, this can also be related to another aside that says: «*I also saw other phenomena related to lightning: how some stars arise; become lightning and cannot leave their new shape.*» (1 Enoch 44:1) Again the detail is recorded in the following chapter: «*And God set them in the expanse of the heavens to give light on the earth, and to rule over the day and over the night, and to separate the light from darkness. And God saw that it was good.*" (Genesis 1:17-18) If the function of the stars is to illuminate the Earth, there is little point in placing them so far away, they should be closer, as with the Sun and the Moon. It is proper to understand that if the function of these The stars is to "master" and "separate" cannot be inert masses, but bodies in activity and movement.

Thus, with this process completed, Yom Rebií (Fourth Day) is decreed. The Paraphrase of Sem also records this process and the hierarchies: «*It was said: "Anasses Duses, you are the infinite Light that was given by the will of the Majesty to establish all the lights of the Holy Spirit in the place, and to separate the mind from the darkness because it was not well, from the light of the Spirit to remain in Hades. Because in your desire for the Spirit arises to contemplate your greatness."* (Shem 11:20-30 paraphrase) This aspect takes us back to when the dark side was limited, but exposes the priority of keeping things that way and locating those who would keep the dark side at bay. While that occurs according to Moses, in consonance with Enoch the following is written: «*And in the fourth interval I ordered that there should be the great luminaries in the circles-spheres of the Heavens. In the circle-sphere, the first, the highest I placed the star-planet Saturn, in the second above I placed Venus, in the third Mars, from the fourth Sun, from the fifth Jupiter, from the sixth Mercury, from the seventh Moon; and in the stars-moons, the small, beautiful-bright ones, those that fly, the minor ones. And watch over the Heavens to light up the Day and over the Moon and over the stars to light up the Night. And the Sun must go around the whole circle [of the constellations] of the zodiac; and 12 circle the zodiac revolve [around] the Moon; and they act according to the meaning of their names and thunder according to the circle of the zodiac that is in front of them, and [the] law of their hours according to their rotation. And it was late and tomorrow was the Fifth Day.*» (2nd Enoch 30:3-7) In the translations into Greek the same source fits as in Hebrew: Kruno corresponds to the Greek Cronos (Saturn, from Hebrew: "Shabetai"), Aphrodit is Aphrodite (Venus, from Hebrew: "Noga "), Aris is Ares (Mars, from Hebrew: "Maedim"), Helios is the name of the Sun (in Hebrew:

"Shemesh"), Zoues is Zeus (Jupiter, from Hebrew: "Tzedek", which means "justice"), Ermis is Hermes (Mercury, from the Hebrew: "kojab", which means "star" or "astro") and finally Selene is the Greek name of the Moon (in Hebrew: "Iareaj").

If this description is equivalent to that of Moses, the luminaries would practically be those 7 bodies, which are described there, in order, as: Saturn -defined as the "highest", that is, it would be the most distant-, Venus, Mars, the Sun, Jupiter, Mercury and the Moon. In addition, there would be the satellites that surround several of these planets. This textual order is inconsistent with any parameter of their distance, location or size, including age. At the same time, in this version, the outermost orbs such as Uranus, Neptune and Pluto are not mentioned, although it explicitly states where the "waters" that were carried above the "firmament" ended up. With many similar aspects we find Inca mythology, where it is said that Viracocha began his work in the world of the ancients (ñawpa pacha) carving in stone the figures of the first two human beings (the first men and women who would become the foundation of his work). These statues are placed by Viracocha in the corresponding places and, as he names them, they animate and come to life in the darkness of the primordial world (ñaupa pacha), because the god had not yet taken care of giving light to the Earth, only illuminated by the radiance of the Titi (a wild and fiery puma that lives on top of the world, surely the jaguar that is intermingled with other animals in the totemic representations of the Inca Empire and previous pre-Inca cultures). When that happened, this world here or Kay Pacha, was still in darkness because Viracocha postponed all his work of erecting a complete world, until the birth of the human beings who were going to enjoy it. Satisfied with the humans, the god continued his project, now putting in their place his children the Sun (Inti), the Moon (Mama

Quilla), and the infinite stars, until covering the entire celestial vault with their lights.

Life Forms

We move on to a new detail about the Earth: «*God said: Let the waters produce living beings, and birds that fly over the earth, in the open expanse of the heavens*». (Genesis 1:20. RVA 60) Catholics, who support evolutionism, would see here a supposed proof that the process from one animal to another coming from water is true, but you have to read between the lines. The deity commands that the water itself produce two things: 1. "shéretz nefesh chayah" (innumerable [forms of] living souls). The word "shrtz" refers to producing innumerable living beings or moving innumerable living beings. More specifically, "shéretz" defines bugs, tiny and innumerable living beings and also insects, as in the case of Leviticus 11:20, when it speaks of "shéretz ha-of" (winged insect). Then we have point 2. "of iofef" (flying flyer). In the current Hebrew language "of" is chicken. The generic definition of "of" identifies a bird, but not by referring to an oviparous or bird directly, but in association with the fact of "flying", so "of iofef" does not translate "aviadora aveadora", because " avear" –something that does not exist in Spanish- would be "fly". So "iofef", when referring to "fly", leaves the word "of" (which makes up "i-of-ef") as the root of "fly" or "flyer". The theory of the evolution of species suggests that birds became such as they came from land animals, not from aquatic creatures. What seems to be seen here is the appearance of two genera of species as the first inhabitants of the Earth, both flying, but he does not talk about that, instead he says that their function was directly to "fly over the Earth" to a specific extent: *"on the face of the Rakiyah of the heavens.* » This could be understood as birds flying in the air in front of the open sky, which would be the atmosphere. In any case, the birds do not fly over the "open expanse of the heavens", since "the skies" are many -not one- and that the complete expanse reaches up to the stratosphere, which is not a zone for birds. No birds have been observed at such heights but only in the lower part of the aerial layer. In other words, the referred layer being a maximum of 80 km and the outermost of the Earth of 1,000 km, the birds fly little over the expanse of the sky.

Defining it in another more extensive way: the atmosphere is made up of several levels starting from the ground, such as the homosphere (75 km), ionosphere (75-420 km), metasphere (420-780 km) and protosphere (780 km to more than 1000). The planes themselves circulate through the field of the homosphere, which has the troposphere (0 to 10 km), the tropopause (10 km) and the lower stratosphere (10 to 20 km), for example. Birds do not fly higher than the air traffic zone, so they are actually moving very low in that "*expanse of the heavens*." Would Moses then speak of the birds? The remains found in the subsoil do not place the birds before the reptiles, and less than the fish, being rather recent. Even with everything, we have not yet arrived at the appearance of these "species". So, what kind of "flyer" would that be that would walk in the same place where the luminaries and the stars that ruled the Earth were? The answer may not yet be revealed, but there is another decidedly significant aside that says: "*And God created the great sea monsters, and every living thing that moves, which the waters produced after their kind, and every winged bird after their kind." species. And God saw that it was good*." (Genesis 1:21) So the Scripture affirms that the deity created the "taninim ha-gdolim". The word "tanin" means "dragon", although it is sometimes translated as "jackal" (although this creature is more commonly known as "tainin"). The word that accompanies it, "ha-gdolím", means "the great ones". But were there small ones? It is safe to say that a Hebrew synonym for "tanin" is "leviathan", and that, furthermore, in ancient tradition, it was practically the same as a "nachash" (serpent), except that the nachash does not have legs (although legends They defined dragons as serpents with legs, adding to their morphology that many were flyers and even fire-breathers, something very common in Far Eastern ideas). With which, to better study the significant aspects of the Bible and what it reviews there, it is good to take the definition of "sea monsters", or modify it, now using "dragon", "leviathan" or "serpent".

That is what the following passage refers to —although let's not forget that before we talked about flyers, then about monsters and again about birds; something to say the least strange. Ezra wrote: «*"On the Fifth Day he commanded him to the seventh part, where the water had gathered, to create living things, birds and fish, and so it was made. The foolish and lifeless water produced living creatures, since it was commanded, so that, therefore, the nations might declare his wondrous deeds." Then you keep in existence two living beings, the one you have called the one Behemoth and the name of the other Leviathan. And you separated from each other, for the seventh part where the water had come together could not hold both. And they gave Behemoth one of the parts that had dried up on the third day, to live in it, where there are thousands of mountains; Leviathan, but he gave you the seventh part, the watery part, and they have remained to be eaten by those you want, and when you want.*» (4th Ezra 6:47-52) The Bible speaks in various parts about Leviathan, Behemoth, the Dragon, the Tortuous Serpent, the Swift Serpent and Rahab, but Enoch refers: «*On that day two monsters will be made to come out separately, one female and another male. The female monster is called Leviathan and lives at the bottom of the [great] sea above the source[s] of the waters. The male monster is called Behemoth, it rests on his chest in the immense desert called Dondain, to the east of the garden inhabited by the elect and the just, where my old father was taken, the seventh since Adam, the first man created by the Lord of the spirits. I begged another angel to reveal to me the power of those monsters, how they were separated in a single day and thrown one to the bottom of the sea and the other to the dry soil of the desert."* (1st Enoch 60:7-9) Later Enoch continues and concludes, saying: « *and the angel of peace who was with me said to me: "These two monsters have been prepared for the great day of God and are fed so that the punishment of the Lord of the spirits will not fall in vain on them, they will make the children die with their mothers and the children with their fathers and then the judgment will take place according to his mercy and his patience."* (1 Enoch 60:24-25)

These strange monsters remind us of the Craken, Typhon –in Greek culture-, Apep –in Egyptian-, Jörmungandr –in Greek- or Sulanut –in Hebrew-: «And when the *Egyptians hid because of the swarms of animals, they locked themselves behind their gates, and the Almighty ordered the Sulanut which was in the sea, to come up and go to Egypt. And she had long arms, ten cubits in length of the man's cubit. And she went up on the roofs and uncovered the roof beams and floor coverings and cut them, and she stretched out her arm into the house and opened the bolts, and opened the houses of Egypt. After that came the swarms of animals inside the houses of Egypt, and the swarms of animals destroyed the Egyptians, and it was extremely severe pain."* (Jasher 80:19-22) It is evident that when speaking of the ocean in Genesis one associates it with water-bearing dinosaurs, but that is not mentioned in the Hebrew version, water is not mentioned at this point. We can thus give rise to those who believe that this is the place of dinosaurs, since they are giant dragons, giant serpents or giant leviathans. By defining them as "the greats" it is because it identifies others who are not, which is why it is also pointed out that it is known what they are talking about. I mean that it does not say: "he created some animals in the likeness of snakes" or "he made large animals that looked like...", but he speaks of them as already known. To this we add that "sea monsters" were not known, if we accept the aquatic dinosaurs that have only been discovered in recent centuries. Not even to a lesser extent the living beings of the seas that are large, such as whales, whale sharks, mantas, giant squids, and others, since this excludes the others. These animals were not known since in Biblical prehistory there were no divers or submarines, nor was there extensive knowledge of underwater zoology. So much so that whales are called "leviathans", because they are huge monsters of the depths. That does not mean that the deity referred exclusively to creating the whales, because they are nothing special, since there are more animals to mention. Although sometimes "tannin" defines a giant sea creature, its translation and valuation start from this order, respectively: dragon, serpent, leviathan, sea monster.

As the oldest cuneiform writing outlined, before the gods the reptiles "came down", but how did they "came down"? Where did they come down from? Did they go down alone or did someone else go down? It does not say "they were created", with which, either they were brought by "someone" -singular or plural- or it is not talking about dinosaurs, but about beings that descended of their own free will. Solomon wrote: «*And the angel ordered the whales (Leviathans) of the sea to come out of the abyss. And let him cast his destiny on earth, and let [destiny] submit [to him], the great demon.*" (Testament of Solomon 1:12) The name of Behemoth passed to the ancient Arab culture as "Bahamut", understood as an aquatic creature - although this figure has undergone enormous and complex modernization processes over time, for which it is currently "Bahamut". "It's something else entirely. It could be said that the original Bahamut was a gigantic fish that resided in a vast sea, being understood as a huge fish that supports the Earth. According to the Arab myth, on its back it supports the weight of a gigantic bull that receives the name of Kujata, which is said to have 400 eyes, 400 noses, 400 mouths, 400 tongues, 400 ears and 400 legs; a large number of appendages between each of which is a distance of 500 years of travel, which gives an abstract idea of the size of the creature, equally abstract. Kujata supports on his back, in turn, a ruby on which an angel rests, who, in turn, supports the 7 hells, which support the Earth where, in turn, on it, are the 7 Heavens. This idea of the heavens is complemented by the legend of the Al-Borak mare, in which it is said that Muhammad would have traveled to the 7 heavens. Although Enoch, Ezra and Job describe Behemoth as inhabiting the deserts or mountains of the Earth, only in Arabia was he seen as an aquatic being. If Behemoth was near the "Garden where the Chosen Dwell", it may not be easy to find, as the "garden" itself appears to have been relocated since the Flood.

So, in all of this, did the birds come out before the fish? Are these descriptions of literal asides about life emerging from our world – not in the process of unfolding from one being to another, but suddenly? In the archaeological and paleontological aspect, this fact can be supported in a certain way: in South Africa, evidence has been found that about 550 million years ago, an enormous sum of multicellular life forms appeared completely suddenly, fully formed and even with a skeleton or shell (exoskeleton). For this magical situation there was no precedent or any adaptive development of previous creatures. Simply, overnight, the ocean was populated with innumerable already perfectly formed species. On the other hand, the deity would not have finished with the work of the water, so it follows that these dragons came out after the living souls in swarms and the flying creatures, which is not specified if they were insects. Apart from these dragons, other forms of life were also created, identified as "nefesh chayah" (living soul). How's that for "living soul"? Are there "non-living" souls? And why does it say "souls" and not "bodies"? The definition of soul, outside the Latin conception (anima), is understood in Hebrew as "identity", a being with its own independent personality, as well as a rational creature. That reasoning was the one that Adam also received, being that reason why he is called a *"living soul"* (1 Corinthians 15:45). So, what difference did Adam have, compared to Jesus, if they also defined other beings and species in this way? Let's see that the previous time they mention the "living souls" is about those that come out in swarms of water, and the next time it is already from the earth itself (there are insects that come out of bacteria, dirt, humidity, water and similar things, it could be theorized that in some way it is possible that many appeared, there would be gaps in this regard to consider this the origin of the bugs).

It is strange, on the other hand, that they could then refer to fish, since their "identity" is different from other animals. Perhaps therein lies the point: each new "design" that appears is set in its respective environment and is distinct from one another, organized into its genres and having its own special qualities. Ironically, in the symbolic sense, men are called "animals" -and precisely the ancient world is said to have been populated, first by sky people (flyers), then the first civilizations appeared on continent-islands (seas) and by last the towns on the mainland. But assuming you're not talking about people in a spiritual vocabulary, it's still hard to support the idea that birds came first before fish and land creatures. Something funny can be seen now, in this verse, where, after the tannin and the corresponding nefesh chayah, he commands to create " *col-of canaf*", that is, "every winged bird". Therefore, which flyers were the previous ones? It says "all", which does not include those previously defined. At the same time, he defines them as "winged", and can a bird not be winged? It is one thing for it not to be able to fly, like an ostrich or a chicken, but it is quite another for it not to have a wing, since that is precisely characteristic of a bird. There are never more or less words or letters in the Scriptures, since that makes them reliable and perfect for investigation, just like judicial laws. So indeed the birds appeared after those water dragons, but those that were behind them do not seem to be birds, since it was not stated at any time that they were "winged", only only that they moved through the skies.

Then Moses reported: «*And God blessed them, saying: Be fruitful and multiply, and fill the waters in the seas, and let the birds multiply on the earth. And the evening and the morning were the fifth day.*" (Genesis 1:22-23. RVA 60) The deity gives the order to these creatures to "multiply", to which anyone would suppose that the animals obey, not their procreative instincts, but the divine call. The objective would be to multiply, but in what sense should they "bear fruit"? It is said that Elohim spoke to Abraham about Isaac (Genesis 17:20), telling him that he would make him "bear fruit" and "multiply" -two different things-, since to multiply is to have great offspring, but to be fruitful is that everything he does It turns out well, growing in blessings, properties, recognition, greatness, etc. These words are similar to those spoken by Isaac to Jacob (Genesis 28:3), wishing him well. At another time, years later, and continuing with this lineage, Joseph calls one of his offspring Ephraim (the second son given to him by Asenat in Egypt), stating that the aforementioned name came because of Joseph's prosperity in Egypt: "*God He made me fruitful in the land of my affliction.*" (Genesis 41:52) Someone might suppose that God blesses his creation so that it develops smoothly and life teems with abundance and beauty, but it is still strange that Elohim would speak to those creatures leaving that perpetual decree and they would always follow it, like remembering parents to children. However, it is not that God spoke to them but that he gives an order or establishes a parameter. Although, why does he speak of filling "the waters of the seas"? Saying "the waters" or "the seas" is unnecessary. We could deduce that there being two seas (those above, on the sky, and those below, on Earth), it was explicit about those that are on this planet, although it says "seas" in the plural, not "sea", and we have already given that there are at least Two Seas, two groups of considerable masses of water called "seas".

Genesis 1:24, following the 1960 Reina Valera translation as a reference, mentions: « *Then God said: Let the earth produce living creatures according to their kind, beasts and serpents and animals of the earth according to their kind. And it was so.*» Once the sea created aerial and aquatic life forms, the Earth would produce its own. The evolutionary theory of species does not fit here either, which points to the appearance of reptiles from sea beings, and terrestrial reptiles would give rise to transitory forms that we call "mammals" –another complete madness considering that lizards lay eggs and mammals give birth (there is no reference to any species that passed from one category to the other, as we pointed out in our book "Creation vs. Evolution"). While verse 20 spoke of the "production" of life forms, under the Hebrew "shretz", saying "shratzu" (populate, swarm, overflow, abound, fill) in verse 24 it speaks of "going out". The translator put the same word, "produce", but here it appears "totzé" (come out or arise). In the case of water, it seems to order the generation of enormous amounts of life forms, while on Earth it orders that the surface itself generate, or come out of it, "nefesh chayah" (living souls). These creatures must have their own typologies or species, as in the case of the plants and trees in verse 11 and the beings of the "sea" in verse 22. It seems then to follow a specific pattern of production of different species and varieties of living beings. However, the description he makes is strange, since we know basically 5 animal types: bird, fish, amphibian, reptile and mammal. Reptiles, amphibians, and mammals would then be missing from the story—although the frogs could be included with the reptiles, assuming the story is not necessarily entirely concise. Even so, he speaks of "behemá" (beast), "remesh" (crawling) and "jaitó-Eretz" (earth souls). Snakes and cattle have been included in certain translations, because it is understood that the snake crawls its entire body, but it is a question of "slithering" not "snaking".

In relation to this fifth period we see the Jubilee again, where it is stated: «*And on the fifth day he created great monsters of the sea in the depths of the waters, for these were the first things of the flesh that were created by his hands, the fish and everything that moves in the waters, and everything that zvuv (flies), birds and all their kinds. And the sun rose above them to prosper (them) and, over all that was on the earth, all that sprouted from the earth, and all fruit trees, and all flesh. These three types He created the fifth day.*» (Jubilees 2:11-13) It says here that the "big monsters" were not all over the ocean, but " *in the depths* ", somewhat strange, since the largest known fish or aquatic mammals do not swim in shallows. bottoms but near sea level, especially the largest cetaceans, since being mammals they must come out every so often to get air. Nor are they considered to be beings from a world below, like sea demons or something like that, since these living forms are defined as " *first things of the flesh* " that God created. But it is described as distinct from "fish," just as birds here are distinguished from "zvuvím" (flying), which primarily translates as "flies." In other words, if the description given is simple, insects were made before the great aquatic animals, and then thirdly birds would appear and fourthly terrestrial animals. Thus, in that Fifth period the creation of said "three types" of existing forms was considered. If at least the beginning of that Era was the basis for the birth of insects, aquatic beings and birds, would it be the age before the dinosaurs? Since by definition dinosaurs are the giant reptiles of prehistory that walked on land, not in the sea or in the air (all in their context are well called "prehistoric animals").

Now, after birds and fish, mammals and reptiles appear, as the story seems to refer, not yet being critical of this, or not at all. He then says: "*And God made the beasts of the earth after their kinds, and cattle after their kinds, and every beast that creeps on the ground after his kind. And God saw that it was good."* (Genesis 1:25) What is now striking is the word incorporated in Hebrew, "Adamáh" (ADMH), instead of the typical "Aretz" (ARTZ.). Why two names for the Earth? Why this new definition? Adamah firstly means "humanity", since it is an extension of "Adam" (man), and let us remember that it has to do with the spirit or influential entity in the union of polarities, which is known as Eros. So, if man did not yet exist, why was he called Adamah? Is it possible that the adamic man was already appearing at that time? « *For thus says the Lord, who created the heavens; he is God, the one who formed the earth, the one who made it and composed it; he created it not in vain, to be inhabited he created it: I am the Lord, and there is no other."* (Isaiah 45:18) In the case of the scribe Baruch, it is stated: « *But you know exactly what you have done through your servants, because we are not capable of understanding what is good as art, our Creator. But once again I will speak in your presence, O Jehovah, my Lord. When old there was no world with its inhabitants, which made it conceive and speak with a word, and immediately the works of creation went ahead. And he told him that you wanted to make man for his world as the administrator of his works, that perhaps it is known that he was not in fact in account of the world, but the world because of him. And now I see that, as for the world that was made on account of us, Oh! It is fair, but we, because of which it has been dictated, turn away.* » (2nd Baruch 14:15-19)

The Earth Human

« *Then God said: Let us make man in our image, according to our likeness; and rule over the fish of the sea, over the birds of the heavens, over the beasts, over all the earth, and over every animal that creeps on the earth.*" (Genesis 1:26) There are many myths about the appearance of man, defined in different ways. Although I will not deal with this matter in great detail in this book –because it will be a collection of the next work- I will clarify the letters of Moses. Comment here that Elohim says "let us make man", but to whom does he say it? And furthermore, what was meant by "man", such that just mentioning it already meant what was going to be created and what it would be like? If we consider that we are in "nothing" and we want to create something, would we give the name of what we are going to do, since it has never existed, or do we first think of the design and idea of what we want to do, and then give it a name? If I were a god and today I said: "I'm going to create a 'tret'" –an invented word, meaningless-, the first thing anyone would ask themselves would be, what is that? Unless everyone already knew in advance that it is a "tret". Elohim does not say "we create a being similar to us", but already speaks of something that has a name, identity and personification, specifically known as ADAM. In fact, the Earth itself begins to define itself just before as "adamic", that is, "humanity". They do not say to create a creature, but a "man like them", that is, man already existed, and what they wanted was to design or form it in their own way. In other words, it is as if they had said "let's make a human, but like us" or "we create an Adam based on our own mould, not like the others". They already knew what Adam was and that prototype already existed previously, only that these "Elohim" decide to make one, according to the textual words, "ba-tzelmenu" (based on our image) and "quidmutenu" (in our likeness, in our image). our relationship). Was there another Adam who was not like them? Paul told the Greeks about God: "*And of one blood hath he made all the races of men, that they may dwell on all the face of the earth; and he has fixed for them the order of the times, and the limits of their habitation ...*" (Acts 17:26)

Monotheism regularly applies mostly theological and dogmatic customs interpreted literally from translations, which says that "God made man in his image and likeness." Now, if God is a spirit, what image does he have? If He is immortal, irrepressible, inexpressible, incomprehensible, limitless, immeasurable, how are we like Him in any of these facets? Does God have eyes, nose, mouth, arms, legs, sexual organs, hair, ears, etc., like humans? And if he also made the woman, and she is his likeness, based on what likeness of God were his female organs created? And if the woman came out of the man, did the first man have male and female genitalia at the same time? We are clearly God's creation, but if we are his "image" and "likeness", then that "God" who created us is exactly identical to us, since the idea that he was the biblical God "neither the heavens of the heavens can contain him", "that he is invisible", "that no one can see", that "he is not a man", etc. God, the Father, has no image, because images are visible figures that can be imitated, distinguished, and materialized. The Heavenly Father is "pure light that no eye can see", it is not a design of a human body between 1.65 and 1.90 m tall. Ultimately, whoever created that Adam had to be exactly like Adam himself.

Another matter is: why, if it is One, does it say: "let's do"? The trinitarian current would say that there were "God the Father, God the Son and God the Holy Spirit". As I have said, the Father cannot be, nor be there, because he is immeasurable and the Earth itself is *his "footstool."* (Isaiah 66:1). For his part, the Son, would he have been part of that before he was born? If he had not yet incarnated (within the flesh) physical, he would not be, and since he was not physical, how is it that Adam does come out physical? If Christ is male, where did the design of the woman come from, especially with her own biological characteristics for the conception of life? Now, if it were the Holy Spirit, then is the Holy Spirit human, is it carnal? Are we identical to the Holy Spirit? If none of the three is physical -since Christ is born in physicality millennia later in Bethlehem-, more than one person created the human under his own characteristics. Now, why should that human be "like" them? Especially if later, seeing him be similar to them, they are surprised and expel him from the Garden. Wasn't that his interest, that he was like them? This is very strange. If they wanted the man to be like them, why are they scared when they see him like them? If so, what was the intention with that? It is as if in the first instance it seemed like a good idea to create a being equal to them, but then, seeing what that implied, they felt jealous, surpassed or vulnerable. If God's special quality is immortality, how did they not expect man to be immortal beforehand? As it turns out, man was immortal at one point, but then they wished he weren't.

He then says that the function of this human would be "ierdú" on certain forms of life. What is striking here is that this is translated as "rule", "have power", "have dominion" or "reign", depending on the publishing house of the Bible. However, later, in other places, the same word is translated as "come down". So what does this mean? In Genesis, chapters 12:10 and 24:16, it is translated as "descend"; in Isaiah 42:102, as "to sail on the sea"; in Genesis 11:53 it is used to refer to Jehovah when he "came down" to see the Tower of Babel; in Genesis 37:35 it speaks of "descending to Sheol"; in Deuteronomy 20:20 and 28:52 he speaks of "surrendering", "falling a city"; then in Judges 15:8 you can see " *and he left* " or "*and he came down.*" Possibly this is the case of quotes like 2 Kings 2:2, in which apparently this verb also has the puzzling sense of "go up" when its meaning is "descend". Elsewhere it is seen as "bring down", "to bring down" (Deut 21:4); "disarm", "dismantle" (Numbers 1:51); "cast down" (Psalm 56:8/7); "to bring down," "to fall" (Ezekiel 34:26); "shed tears" (Lamentations 2:18); "subdue the peoples" (2 Samuel 22:48). And equally it applies to Genesis 39:1 when it speaks of "being brought down" (Genesis 39:1), or "being taken apart" (Numbers 10:17) or "being thrown down" (Zechariah 10:11). Ultimately, Adam, should he descend or make others descend? The text says that this was applied "bi-degat ha-iam" (in the fish of the sea). But, what other fish would it be, if the fish is in the water? There is no room for redundancies here. That definition of "dagat" or "deget" in association with "fish" is rarely used (Genesis 1:28 and Ezekiel 29:4-5), which makes this a likely reference to something more relevant than just a fish.

The Canaanite culture had Dagon as its main deity, from whose name comes the voice "draco" and then "dragon". This same word comes from the Akkadian and Aramaic "dag", which is "fish". The designation can identify a being with scaly skin and similar or inhabitant of the ocean depths, but it was the way of referring to the fish gods that were seen thousands of years ago in southern Mesopotamia, the first gods to appear, according to of that region. It should be noted that until now the definition of what we call "fish" had not been used, which anyone would assume refers to the living souls of the seas, which appeared in previous episodes. In any case, it is rare that they say that Adam had to "descend into the fish", which is the most accurate literal translation. The other thing would be to apply the less common idea, which would be the fact of "subduing" the fish. And how is that done? It is clearly understood that Adam receives a pre-eminence and superior status to the other forms of life on Earth, but it must be recognized that it is strange to use this word to refer to it. In addition, that "yerdú" of Adam should also be about the "of ha-shamaim" (flyers of the sky), about the "behemá (beasts), «*and in all the Earth and in every creeping thing that crawls on the Earth.*» Clearly his goal was to be above what was in the sky, in the seas, on Earth and that he could hold sovereignty over the inhabited world, not like Tarzan but with an authority that was given to him. This does not seem like a divine interest in creating a being to delight in watching or worshiping, but rather in controlling existing things, to the point of "subduing" them and "descending" into or over them.

This is how the classic phrase comes from: «*And God created man in his own image, in the image of God he created him; male and female he created them."* (Genesis 1:27) The Hebrew phrase returns here to identify man as something of which there was already evidence, not calling him "Adam", but referring to him as "ha-Adam" (the Adam), the one they had created, emphasizing that they created it in accordance with their own image, countenance, kinship or visual form. And he adds: *"ba-tzelem Elohim bará oto...",* which translates: *"in the image of the gods he created him."* This process being a job to make accessories: "masculine and feminine". In other words, the work of creating this type of human consisted of making them of both sexes, evidently for procreation purposes, since, if his objective was to dominate, and it would be eternal, he could do it alone or with a friend of the same characteristics. -but not pointed out here the sexual aspect of masculinity and femininity. So when they appear they tell them what their function is and command them to be fruitful in this and to have offspring (verse 28). The following is: «*And God said: Behold, I have given you every plant that bears seed, which is on all the earth, and every tree in which there is fruit and that bears seed; they shall be for you to eat."* (Chap. 1:29) They make it clear that the plants were placed there for their food, so it seemed that their appearance was deliberate, since Elohim did not make the plants for himself, but for that man. He points out their food of products with seeds that in turn produce other seeds and trees that bear fruit which generate other trees in turn. Now comes another punctual aspect: «*And to every beast of the earth, and to all the birds of the heavens, and to everything that creeps on the earth, in which there is life, every green plant will be for food. And it was like that."* (Ch. 1:30) All eating grass? Birds do not eat grass, nor "carnivorous" mammals, nor reptiles. Precisely their metabolism was created differently so that they ate differently, which is why carnivores have fangs and it is appreciated that they do not eat grass. A fundamentalist Christian would say that clearly at that time there was no savagery, no cannibalism, no death –food chain-, but then man was not allowed to eat meat either, but rather fruits, vegetables and seeds. Why did mankind start eating meat? Wasn't what the Earth produced enough? Would there be shortages? Well, here the work of 6 yomim concludes: «*And God saw everything that he had made, and behold, it was very good. And the evening and the morning were the sixth day.*" (Genesis 1:31)

That interval known with the number 6 is linked to man: « *And on the sixth day He created all the animals of the earth, and all the cattle, and everything that moves on the earth. And after all this he created man, a man and a woman he created them, and gave him dominion over everything that is on the earth, and in the seas, and over everything that flies, and more beasts and more cattle, and over everything that moves on the earth, and over all the earth, and more than all this has given him dominion. And these four guys that he created on the sixth day. And there were a total of 22 guys. He and all his work finished on the sixth day - what is in the heavens and on the earth, and in the sea and in the abysses, and in light and in darkness, and in everything. And He gave us a great sign, the Sabbath, that we must work 6 days, except Saturday, abstain on the seventh day from all work. And all the angels of the presence, and all the angels of sanctification, these two great classes - He has commanded us to keep the Sabbath with him in Heaven and on Earth.*" (Jubilees 2:14-20) Moses' reference is similar to that of the later scribe who recorded these facts: "*On the sixth day he commanded him to land before driving out cattle, wild animals, and creeping things; and more of those who place Adam, as ruler of all the works you have done, and of what we have all come to, the people you have chosen. "All this I have spoken before you, Lord, because you have said that it was for us that created this world. As for the other nations that are descendants of Adam, he has said that they are nothing, and that they are like spittle, and have their abundance compared to a bucket drop. And now, Lord, these nations, which are like the fame of being nothing, rule over us and devour us. However, your people, who have asked for their firstborn, only begotten, so that you are jealous, and dearest, have given themselves into your hands. If the world has been created for us, why not have our world as an inheritance? How long will this be like this?* » (4th Ezra 6:53-59)

In previous chapters of this 2 revelation that I had, Ezra, speaking to the Most High, said to him: «*"O sovereign Lord, did you not speak at the beginning when you planted the earth - and that you [did] without help - and commanded the residue and that gave Adam a lifeless body? Yet it was the creation of his hands, and he breathed into him the breath of life, and he was made alive in his presence. And he led him into the garden that his right hand they had planted before [the] earth appeared."* (4th Ezra 3:4-6) It is said again that the Garden existed before civilization and the very beginning of the foundations of it. Then man would come and fill the Earth: «*"That is why I have given them women so that they can fertilize them and engender children for them and so that they do not lack on the earth."*» (1st Enoch 15:5) Going to the description of Enoch we see how he concluded his story about Creation by talking about the same things as the others who came after him, but emphasizing specific details, such as the consistency of man, not of "dust", but of existing elements. It is also taught that 4 great divine watchmen guide you. It is also expressed that he was made aware of the good and bad things and that he decided, by himself, which path he wanted to follow. However, God told Enoch that He himself knew perfectly well his project -it was not an idea without a basis-, but He did not want to tell Adam, lest this hinder Adam's life and his destiny, causing him to stumble and transgress:

«And on the Fifth Day I ordered the Sea to bring out fishes and various flyers (birds) multiplying and all dragged that crawls on the Earth and going on 4 on the Earth and they fly in the male

and female wind-spirit in the midst of them, and every soul that breaks-grieves all life. And it was late, and the Sixth Day was tomorrow. And on the Sixth Day I ordered my Wisdom to create man starting from 6 bases: his flesh from Adamá; his blood from the dew; and sun his eyes from the abyss of the sea; his bones from the stones; what he thinks of the speed [of] the malachim and the densities-clouds their tendons-ligaments; and his hair from the grass of the Adamá; his soul of my breath and my spirit. And I gave him 7 qualities, the 7 for the flesh: sight for the eyes; the smell for the soul; the joy for the ligaments-tendons; taste for blood, patience-tolerance to bones; meekness-tranquility for thought. And I thought why I said-call Word wise because of what exists that is not seen and is seen I made man, both dead and living, and the image knows Word and there is no in all Creation like him: small in greatness and in the big littleness. And I put him as the second custodian-ordained malaj on Earth, upright and great, and respected. And I seated him as King of the Earth and there was no one like him among my wise men. And there was no equal to him on Earth of all my creations. And I named it the 4 wind-spirits: from the East, from the West, from the North [and] from the South. And I put in custody 4 stars of the cooled-dry ones and called his name Adam. And I showed the dispositions and he looked at the two paths: light and darkness, and I told them: That is going well and that is bad to go, [I must] know if he has discernment towards me [or] with hatred towards the excellent direction, [and who] of his seed loves me. And I saw my project but he did not know about the project, and if he knows he will do it badly, sin and all that is his will be sin, and I said: after sin there is no Word but death. And I ordained him a tabernacle and [deep] meditation came upon him and he slept. And I took his rib and marked it and created him a woman. To address him death at the hand of his wife. And I took the last letter of his name and called his name, they were him: Adam, [and] they [are] humanity and the living." (2 Enoch 30:8-16)

WE WILL CONTINUE IN our next book: "The Rebellion of Sakla II, The Serpent".

God bless you!

Don't miss out!

Visit the website below and you can sign up to receive emails whenever Frederick Guttmann publishes a new book. There's no charge and no obligation.

https://books2read.com/r/B-A-DKUGB-XZEBD

BOOKS 2 READ

Connecting independent readers to independent writers.

About the Author

Israeli writer, researcher, disseminator, documentary filmmaker and influencer. He is the writer of more than 35 books, mostly research and dissemination theses.

Read more at https://www.frederickguttmann.com.